D1601179

A Naturalist's Scrapbook

BY THOMAS BARBOUR

(*Continued from front flap*)

Ever since his first visit to the Harvard University Museum at the age of twelve, Thomas Barbour wanted to be the director of Harvard's great museum. He spent a lifetime in Agassiz's chair behind Agassiz's desk, and his purpose was always to keep the Harvard museum the greatest university institution of its kind in the world. During his nearly 40 years at Harvard, Dr. Barbour traveled to some of the remotest parts of the earth. His almost annual explorations took him to India, Burma, China, Japan, the East and West Indies, and Central and South America. From 1927 until his death he was custodian of the Harvard botanical garden at Soledad, Cuba and for 23 years officer in charge of the laboratory at Barro Colorado Island in Gatun Lake, Panama Canal Zone. His bibliography of scientific publications fills two small pamphlets, and he published three books for popular reading, among them *Naturalist at Large*. A NATURALIST'S SCRAPBOOK is just what it claims to be, a scrapbook—the scrapbook of a many-sided and busy man who found time in the course of an extremely full life not only to run a university museum and to raise a family, but also to write, hunt, fish, and collect books.

A NATURALIST'S SCRAPBOOK

LONDON : GEOFFREY CUMBERLEGE
OXFORD UNIVERSITY PRESS

THE AUTHOR

A Naturalist's Scrapbook

THOMAS BARBOUR

ILLUSTRATED

Cambridge
Harvard University Press
1946

QH
31
.B215
A3
1946

COPYRIGHT, 1946
BY THE PRESIDENT AND FELLOWS OF HARVARD COLLEGE

Second printing

LIBRARY
LOS ANGELES COUNTY MUSEUM
EXPOSITION PARK

8/15/46 Vroman's

PRINTED AT THE HARVARD UNIVERSITY PRINTING OFFICE
CAMBRIDGE, MASSACHUSETTS, U. S. A.

Preface

I HAVE always been a pack rat, a frank and unashamed pack rat just for the simple reason that I enjoyed being one. My dear friend, Professor Edward S. Morse, who was of the same breed, rationalized all sorts of benefits which accrued to mind and soul as a result of the collecting instinct. This may be possible. He believed in teaching children to collect cigar bands, match boxes, and Heaven knows what all. I never descended quite this low, but a pack rat I have been through my whole unashamed life.

Here are some little essays which I have accumulated which describe all sorts of notions and events. I hope reading them may give some of my friends as much pleasure as writing them has given me.

Cambridge, Massachusetts — T. B.
October, 1945

Contents

Illustrations

A NATURALIST'S SCRAPBOOK

CHAPTER I

The Roving Eye

ONE of the most important functions of a museum director is to bring about the transfer of material which is scientifically important from unappreciative to appreciative, and *safe*, ownership. I remember worrying for years over the fact that the beautiful type of a fossil sea-turtle, *Carolinochelys wilsoni*, was in private hands in Charleston – treasured by a lovely soul, but in a house which was by no means fireproof. After his death I communicated with his son and he realized the importance of the arguments which I presented to him. The collection containing this priceless type went to Cambridge, where it is now safe.

There are probably few of the old houses in Charleston where fossils from the phosphate beds are not to be found. In the heyday before the Florida phosphate fields came into production almost everyone who had land in the vicinity of Charleston mined phosphate rock on a more or less considerable scale. It is extremely unfortunate, though quite understandable, that the reverence for the past which permeates Charleston's being should make most of the possessors of these fossils reluctant to part with them, and it is a

LIBRARY
LOS ANGELES COUNTY MUSEUM
EXPOSITION PARK

tragedy that this material should be prevented for these purely sentimental reasons from playing a significant part in the increase of human knowledge. They would play such a part if they were turned over to the care of an institution where vertebrate palaeontology is a discipline which is actively cultivated. I know that there is fine material, even including some very significant type specimens, which is and has been for many, many years in private hands. Fortunately, material of this nature is unlikely to deteriorate in the ordinary sense of the word as a collection of insects or birds might do; but specimens fall off the shelf to the ground and crack or are stepped upon, and there is no possible disputing the fact that they are not safe until they reach the storage cases of a research museum.

Opportunity to make a transfer of ownership beneficial to the Museum of Comparative Zoölogy as well as to the original owner of the specimen involved has offered itself to me by chance upon several occasions. Once, on the way to join my brother, who was salmon fishing at Matapedia, I stopped to examine the rather somnolent overcrowded collection of mounted birds in the Redpath Museum of McGill University. Standing back in the darkness on an overcrowded shelf there was a Wood-rail with a heavy and peculiarly shaped bill. I couldn't read the label clearly, but saw enough to convince me that the words upon it were "Habroptila wallacei." This was an avian treasure. I sought out Dr. McBride, at that time head of the Department of Zoölogy in the University, and told him I would like that Rail, since a colleague of mine was at that time studying the bills of birds in relation to their feeding habits, and that I would return to the Redpath Museum a specimen more spectacular for purposes of exhibition. The facts as stated were true and the Rail was sent to Cambridge. It turned out to be one of Wallace's actual cotypes, one of the three specimens which he secured on Celebes from which the species had been described. This was a specimen which had been completely lost from sight, one which was serving no useful purpose where it was and which was indeed treasured when it reached its destination. I may add that the bird was not collected again until a few years ago, and I believe that then only two more specimens were taken.

Some years later the American Ornithologists' Union decided to hold its annual convention in Quebec. Fairly sure that there was a museum at Laval University, where the meetings were to be held, I went to Quebec a day or two before the rest of the members, to have a chance to examine that museum at my leisure. I saw there the shell of an enormous turtle labeled "Testudo elephantopus, Galapagos Islands." A glance suggested to me that the specimen was wrongly labeled and that it was something else altogether; I suspected it to be a giant example of *Testudo ponderosa* from one of the islands in the

Indian Ocean. I sought out the custodian and told him that there seemed to me some doubt about the identification and the accuracy of the locality, and that I would send him a Galapagos tortoise with definite data as to locality, well mounted and suitable for exhibit, if he would send the other old shell to Cambridge. When it reached us here, I found inside the shell, writen in ink, the number "1937" in unmistakable French script, with a strong stroke across the "7." On the chance that I might get some information, I wrote my friend Dr. Paul Chabanaud of the Jardin des Plantes in Paris and asked him what the register showed under this number. He replied promptly, "This specimen is not now in our collection. It was sent many years ago to Laval University in Quebec. It came from one of the Aldabra Islands in the Indian Ocean." I was very happy to receive this news, and studying the specimen carefully, determined definitely that it was indeed *Testudo ponderosa*. I believe that this is the same species as an enormous specimen which a few years ago was living in the garden of Government House in St. Helena. This is said to be the only creature on the island which was living when Napoleon was there in exile. I have some excellent photographs of my daughters feeding it, as it stretched forth its long skinny neck and stood high up on its big stumpy legs. It was indeed the living image of the tortoise I got in Quebec.

Some years ago I visited North Carolina's State

Museum, where perhaps my earliest scientific correspondent and old friend, Mr. C. S. Brimley, is the Director. I mentioned to him that we were extremely anxious to have in our museum a North Carolina beaver. We talked of an exchange and settled upon one which was mutually satisfactory. Some time later, when the beavers arrived in Cambridge, what was my delight to find that the specimens sent were actually cotypes of the species. Inasmuch as the hides were beginning to become what we call "grease-burned," it is well that we got them when we did, for I believe now they have been made safe for all time. The specimens which we sent to Raleigh were beautifully mounted examples of an extinct species of bird, and I had supposed that the beavers which I secured also represented an extinct species. My surprise may be imagined when later I learned that a few individuals still existed around the headwaters of the Flint River in Georgia. With the permission of the state government, I had a pair collected from this locality, so that while the Carolina beavers were not representatives of an extinct form, as I had believed they were, the fact that they were cotypes of *Castor carolinensis* more than made up for the fact that they represented a living, not an extinct species.

A marvelous story of what may happen to a collection which by chance has found the wrong repository is told by my colleague Merritt L. Fernald, the well-known botanist. The story of the Chester

Dewey Herbarium, of Carices he writes, "containing 97 types of species and varieties proposed by him, is typical of that of many small collections which, after the death of their assembler, fall into unappreciative hands. A loyal graduate of Williams College to which he had consecrated his early years of teaching, at his death in 1868 Dewey bequeathed his very precious herbarium and remarkable library to the College." The Catalogue of Williams College from 1868–69 at least until 1876–77 carried a mention of Dewey's collection.

> In April, 1900 [continues Mr. Fernald], needing to study certain of Dewey's types, I wrote to Williamstown to inquire if I might there examine them. The reply from the late Professor Samuel F. Clarke was discouraging:
>
> "April 5, 1900. . . . I have no recollection of having seen the 'Dewey collection.' . . . Professor Chadbourne also had something of a collection, which was later removed from here, some 12 or 15 years ago. I do not know where it is."
>
> As if further answering my query, I very soon received from the late William C. Strong, professor of physics and other sciences at Bates College in Lewiston, Maine, a letter, stating that in a wooden case in an attic of the Science Building there was an old collection of "grasses" made by a man named Dewey. They were poor specimens and of no interest (to a physicist), so they were going to burn

them up. Suddenly remembering my interest in such unpractical things, he was writing to ask if I would like to have some of them!

I hastened to Lewiston. There, in pigeon-holes in the attic, was Dewey's herbarium, the specimens fastened by exceedingly long pins run through the labels, to sheets of coarse gray paper. Since the sheets had originally been too large for the pigeon-holes, the so-called curator had pruned them down with large shears, indiscriminately chopping specimens and labels. They could then be rammed into the pigeon-holes. There the responsibility ended. Now, the pigeon-holes being needed for something worth while, a bonfire of the useless 'grasses' was next in order. I dined with the College President and explained to him that, although these specimens were not wanted by his institution, it would be most unfortunate to destroy them; in the hands of a specialist on the sedges they would be of greatest scientific importance. My diplomacy was too open, for I promptly received the reply: 'If these Dewey plants would be of value to Harvard University, they are certainly of value to Bates College.' The bonfire was immediately abandoned.

Four years later, at the beginning of a new regime at the College, the question was reopened, the exchange proposed was accepted, and in February, 1904, I returned to Lewiston, to pack the collection for shipment to Cambridge. Pasteboards around some of the packages bore, in the characteristic hand of Sereno Watson (former Curator of the Gray Herbarium) the address, "President Paul Ansel Chadbourne, Williams College." Since Watson's remarkable botanical career began with the CLARENCE KING Exploring Expedition and he had come to Cambridge several years after the death of Chester Dewey, it was evident that I had stumbled upon a first-class mystery.

Briefly, the plot worked out as follows. Watson brought to Cambridge, to work up with the aid of Asa Gray, the vast collections of himself and William H. Brewer secured on the Clarence King Expedition. From this resulted the *Botany of California* (1876–80). William Boott of Medford treated Carex in that work. Consequently, since it was desirable to see Dewey's material in the genus, this was borrowed; and after Boott had finished with it, Watson returned it to Williamstown, addressed to Chadbourne, himself an amateur naturalist, whose considerable herbarium had also been deposited at Williams College.

In the *American Naturalist*, 10: 370, for June, 1876, there appeared the following item:

"AN HERBARIUM FOR SALE. – An herbarium containing specimens illustrating six thousand species of plants, is offered for sale. Full particulars can be obtained from President Chadbourne, of Williams College, Williamstown, Mass."

Seeing this, a gentleman in Lewiston bought Chadbourne's herbarium for Bates College, Dewey's herbarium arriving with it.

The sequel is short. Receiving a request for the loan of the type of Carex barbarae of Dewey, I went to the portion of his collection where it should be found. There was nothing in the cover. Examination of the organized Gray Herbarium, however, revealed it, properly mounted and bearing the printed presentation-slip, "ex Herb. William Boott." William Boott had not returned the Californian material of Dewey; but in the end the wanderers and the original collection were reunited. When I reported the incidents to Williams College I received a gratifying letter from Professor Clarke, stating that they were glad that the runaway collection was now safe in a haven where it would be cherished and constantly consulted.

This is as amusing a history as I have ever read and I hope I may be forgiven for reprinting it from the Proceedings * of the American Philosophical Society. It certainly shows that eternal vigilance pays, for one never can tell where treasures to enrich the collections entrusted to our care may be found. Ethnological articles of the greatest value, ranging from priceless feather capes brought from Hawaii a hundred years ago to lovely baskets made by the Cali-

* Volume 86, No. 1, 1942.

fornian Indians at a period when their unrivaled workmanship was at its height, or Fijian war clubs brought back a hundred years ago, have been found again and again in the attics of old New England houses, and the end is not yet.

For a number of years I have been a veritable mendicant friar in my approaches to the various companies mining phosphate rock in Pierce County in Florida. The years of my mendicancy date from 1916, when, spurred on by Doctor Sellards' description of *Tomistoma americana*, I wrote Mr. Anton Schneider, who discovered the type, to ask that he might favor us with subsequent specimens as they appeared. This he did, and the ensuing correspondence gave rise to a number of pleasant acquaintanceships among officials of the various mining companies.

Driving north from Miami to Gainesville in April 1942, Doctor T. E. White and I decided that it would be well to cut up through Pierce County again and see what might have turned up since our last visit. As usual, we picked up a number of treasures here and there. At one of the plants I asked what had become of the material which was on exhibition at the Century of Progress Exposition in New York. It was finally located and packed for shipment to Cambridge. This included one specimen which, so far as I know, is entirely unique. It is a complete lower jaw of a giant whale apparently allied to the finback, a whale formerly not uncommon off our New England coast.

This proved to represent a distinct and hitherto unknown species. It is not without an amusing side to record that this jaw spent a year and a half in New York at a great public exposition unnoticed by any of the innumerable naturalists from various museums who must have visited the exposition time and again and seen it with unseeing eye. The exhibit had been shipped back to Florida, where it lay around for a considerable period of time before making another trip north, finally to be described and named and to fetch up at last in an exhibition case in our Museum.

Years ago the Museum fell heir to a considerable series of reptiles which had once belonged to the Deseret Museum in Utah. I occupied myself for some time identifying the material and preparing it for incorporation into our study collection. All went well until one day I was completely stumped by what appeared to be two chubby little lizards about five or six inches long, marked "Buckskin Mountains, Utah," whence most of this material had in fact been collected. I studied them for some time without an inkling of their identity. Then suddenly I pinched myself for joy and could hardly believe my eyes, for it came to me all of a sudden that these in reality were two young individuals of that extraordinary beast known as the tuatara, or Sphenodon, of New Zealand. Sphenodon is the only living survivor of the group of Rhynchocephalia. One hundred and eighty million years ago, more or less, as we know from

Triassic fossil remains from various parts of the world but especially from southern Brazil and South Africa, giant allies of the tuatara peopled the earth with many genera and species, some of great size. All have long since disappeared. Sphenodon in New Zealand is the sole survivor. It is a reptile which looks like a lizard but is not one, as its skeleton conclusively shows. It was at one time abundant, but being sluggish and inoffensive it disappeared rapidly before the dogs and pigs which came with the increasing settlement of the country. The surviving remnant is preserved on the Poor Knight's Islands off the northeast coast of the North Island, on Stephen Island in Cook Straits, and on the Brothers Islands, also in the straits. On these tiny rocky islets, difficult of access, these reptiles are now strictly protected. Only a small number persist, sharing burrows in the ground with shearwaters, whose young they occasionally eat. Years ago I saw living individuals in the London Zoo, and I have seen preserved specimens in many museums. Once a big collection of the animals was made at the Bay of Plenty and distributed by a well-known natural-history dealer before there was any idea that the animal was going to disappear. But of all of the specimens which I have seen in museums in this country and abroad, I have never seen any but adults, so that, to us, these two young individuals were veritable reptilian diamonds.

The possession of these little treasures by the

Deseret Museum is explained by the fact that for some years Mormon missionaries proselyted extensively in New Zealand. They must have sent these specimens home, where, after their arrival, they were tucked by mistake into a bottle which had once contained material from the Buckskin Mountains and which was still so labeled.

But there is more to the story than I have told, for one of the specimens had had its tail broken off and a new one had grown to replace it, as is commonly the case with reptiles. In all such cases the arrangement of the scales of the rejuvenated portion is always much more simply designed and less ornate than the arrangement of the scales of the tail of an uninjured adult. It has been assumed that these rejuvenated portions in reality represent a reversion to an ancestral type. Here was a chance to test this assumption, for I remembered that in the wonderful Haberlein collection which Mr. Agassiz purchased many years ago, composed of material from the lithographic slates of Solenhofen, there was a fine example of Homoeosaurus, a small Rhynchosaur of the Jurassic Age, say 130 million years ago. This beast was in some degree at least ancestral to our Sphenodon. We possess the obverse and reverse of the fossil, as it was found in a slab of rock which was cracked apart. I wondered at once whether there might by chance be any good impressions which would show the character of scale formation of the reptile on the Solenhofen slabs. To make a long story short, there were distinct and well-preserved impressions of scales along the tail region and these were clearly discernible under a dissecting microscope. When compared they matched the squamation on the rejuvenated tail of our little Sphenodon almost exactly. Thus the case was proved. The scales on the reproduced tail were reversionary and surprisingly similar to an ancestral form, and it was with great pleasure that, in company with my friend Henry Stetson of the Museum staff, I wrote up this story, with some excellent figures, and published it in the Bulletin of the Museum. It is seldom that a chance find turns out to be at once extremely rare and so really significant.

I repeat again, it is the unexpected which provides the spice and savory for the ordinary day-in day-out manner of life of the average museum curator.

I have talked now a good deal concerning problems in addition connected with the Museum. Once in a great while, however, a problem in subtraction has to be reported. One example shall suffice.

Years ago it was a custom of Harvard University to exchange professors with those of foreign countries. I suppose something was gained in this way but, generally speaking, it has always seemed to me that the Harvard teachers did a reasonably adequate job. One jovial and extremely erudite exchange visitor was particularly interested in whales, in their anatomy, and especially in their embryology. He greatly admired several tiny embryo whales, wee mites but a few inches long, which had been in the museum for many years but which nevertheless were still well preserved. Specimens such as these are, as one might expect them to be, extremely rare, greatly prized, and very difficult to secure. As I remember it, there were three or perhaps four of these. The Professor gazed repeatedly at the little embryos with such clear and undisguised admiration that I was always a little nervous about them and I am frank to say that I was not greatly surprised to find that they turned up missing after he left Cambridge to return to his post in Breslau.

I have a feeling that the Professor's conscience bothered him a little, because after he returned to Germany he sent us quite a collection of local European vertebrates of one sort or another, the value of which may have bulked in his own eyes as an adequate return to be made to a Museum *in partibus infidelium.* Senator Lodge proposed quite seriously that the return of these specimens be written into the peace treaty after the first World War. Senator Lodge's politics, however, most unfortunately, were not of such a nature as to bring his opinions to carry much weight with the Administration, and nothing ever happened to set matters right. The Professor now is dead these many years and he never got around to publishing any observations based on his ill-acquired booty.

CHAPTER II

The Museum of Comparative Zoölogy

THE story of the beginning of the Museum of Comparative Zoölogy is still too well remembered to require repetition. Harvard College had possessed collections, not very distinguished to be sure, for nearly one hundred years before the coming of Louis Agassiz in 1847 electrified the museum consciousness of the University and the community. Agassiz brought his collections from Europe and added them to those already here. He was able to acquire material far and wide through the magnetic spell which he cast over everyone with whom he came in contact. Finally he did the impossible, which was to persuade the hard-headed and usually penny-pinching Yankees who composed the membership of the Great and General Court of Massachusetts actually to grant money for the building of the Museum in 1859.

This having been done, he proceeded to beg from his friends, who found it hard to refuse him. The Museum then existed for years as an independent institution, with its own self-perpetuating Board of Trustees. In 1873 came the Professor's death, and the institution became affiliated directly with Harvard

College. The Board of Trustees was transformed into an independent Faculty, recognized by the Corporation of the University as having all the rights and privileges that were possessed by the Faculty of Arts and Sciences, Law, Medicine, or Divinity.

After the Professor's death his son Alexander developed the Calumet & Hecla Copper Mine, thus providing the means which he so generously and constantly used, particularly in the preparation of the unrivaled series of Memoirs and Bulletins published by this Museum. I simply mention these few facts to give continuity to my narrative and to provide the core about which I wrap my own experiences. The history of the origin and development of all three of the museums which I have served during my lifetime, that in Cambridge, that in Boston, and that in Salem, is available in published form to those who may have sufficient curiosity to seek this information.

My first chance visit to the Museum of Comparative Zoölogy is described in my *Naturalist at Large*. I came here to Harvard College as an undergraduate in 1902, and from then until now I believe never a day has passed when I was in Cambridge and not ill, that the day or a part of it has not been spent here in the Museum. I wish mightily that I had lived early enough to listen once to Louis Agassiz lecture and to see him draw on the blackboard. He must have been a charmer in every true sense of the word. He was not all saint by any means — who is? Read the life of

Edward Sylvester Morse and you will find out what I mean. A little leaflet by Henry James Clark, Adjunct Professor of Natural History in the Lawrence Scientific School, published in 1863 and now excessively rare, tells the story in more detail. This leaflet is entitled *A Claim for Scientific Property*, and I have never seen but a single copy, which is in the library of the Boston Society of Natural History. I believe the fact that Professor Agassiz was educated in Germany and unconsciously absorbed a point of view which is common amongst German professors to this day was what caused the trouble. As a matter of fact, Professor Morse made up with his teacher and they were dear friends for many years.

Agassiz possessed the true didactic instinct, if we may call it such. He loved to lecture and to teach and he did it extremely well. On this point everyone agrees. That he was possessed of great charm of manner and was a highly social person, in the best sense of the word, is also agreed upon by everyone. How else would it have been possible to spellbind the members of the Great and General Court of Massachusetts, extracting what in 1857 was an enormous sum of money for the benefit of a museum, of all things!

Louis Agassiz was an extrovert; his son the very reverse. Alexander Agassiz I had the good fortune to know, and Henry Bigelow and I were the only two students of our time, so far as I am aware, who ever made his acquaintance. During the last years of Mr.

Agassiz's life he was held in awe, indeed was considered a terrifying and almost legendary figure by all the graduate students here. It is true that he sternly refused to be bothered when the "out" sign was turned on his door, but remember he was a very busy man. When he turned the sign to "in" no one could be more simple, charming, and delightful. I am firmly convinced that no more methodical person ever lived. He came and went on long expeditions to the far Pacific or to Calumet and back to Cambridge always with exact dates set for arrival and departure. He went regularly to the Calumet and Hecla Mining Company office on Ashburton Place, in Boston, driving in and out in a brougham with a very good-looking pair of horses. You could set your clock by the arrival and departure of his carriage. He came and went with the utmost exactness of schedule. He also worked at home during the early part of the evening, but at ten o'clock, with great regularity, he walked over to the Holly Tree at Harvard Square and there partook of a pair of the famous poached eggs for which the Holly Tree was so well known. I often did the same thing myself, and every once in a while I contrived to meet him and walk home with him. Not infrequently he would ask me into his house, and we would sit gossiping for a while. I think his reputation for irascibility was rather fostered by persons who disliked him and perhaps were jealous of the position which he occupied in the zoölogical world.

Certainly he was the soul of courtesy to me, and I look back with great pleasure upon the fact that I made it a point to make his acquaintance. The whole picture of him which I carry in my mind is one which I am very proud to possess.

Mr. Agassiz disliked lecturing as much as I have always disliked it, and he was very far from being a facile speaker in any sense of the word. I refer my reader at this point to what was said about the two men, father and son, in my *Naturalist at Large*, for I think I made there a very fair comparison and I discussed them after having given the matter a lot of careful thought.

Another person whom I met, thanks to Austin Clark, then a graduate student, a kind friend and now a distinguished curator in the United States National Museum, was Mr. Samuel Garman. He was the Curator of Fishes and Reptiles until the charge was divided, and long afterwards I became Curator of Reptiles and Amphibians. My secretary for many years, Miss H. M. Robinson, at my request has drawn upon her recollections of Samuel Garman so that no trace of the shadow of prejudice may be laid at my door. The truth is that as every great man has his weak spots, Mr. Agassiz's was a very obvious weakness in judging men. Miss Robinson's account follows:

> A glance from a window on the Oxford Street side of the Museum, a little after nine o'clock any morning, would have provided you with the sight

UPPER: CAMBRIDGE IN 1857 OR 1858
LOWER: THE FIRST SECTION OF THE MUSEUM, 1859

THE M.C.Z. FROM THE AIR

of Mr. Samuel Garman approaching the building. He was not particularly tall, and his height was lessened by the stoop-shouldered posture which his eighty years had given him. He came down the walk, looking neither to the right nor to the left, carrying his green cloth bag and seeing no one. He always wore, winter and summer, the same shabby long black overcoat and a black soft hat, and he looked like something which had hung over from the last century, or perhaps a human blackbird. He would enter the Museum, walk hurriedly across the hall, stoop and unlock the mail box which bore his name, and take out his mail. It was here that I usually ran into him and in return to my greeting he would mumble a not unfriendly 'good morning' but would vouchsafe never another word. He would then go out through the east side door, step down into the areaway, where he would unlock his own outside door, and disappear into his laboratory. The curtains were always drawn at his windows, and the only way to gain admission was to ring a pull doorbell beside his locked laboratory door. If he was there and was so inclined he would open the door a crack and inquire who wanted him. I don't think I ever attempted to gain admission. I am sure I never saw the inside of his room while he was alive.

I once met a lady who had been a classmate and friend of his daughter. She told of going to play with little Miss Garman and said how strange she thought it was that they were never allowed to make any noise if Mr. Garman was at home. In fact, when he came home, she was usually sent away so that he would not be disturbed. She said she never saw anything of him except as he came and went. He was never visible about the house.

After Mr. Garman died, his office was opened for the first time to the other members of the Museum staff, for more, that is, than a brief visit. For years after I finally won his favor and worked in his room, when the bell rang and the door was opened and anyone let in the newcomer would find the desk covered with newspapers spread out to conceal whatever work Garman had in progress. This was done because Garman was bitterly afraid that someone might find out what he was dissecting or what otherwise was absorbing his interest and, perhaps, in some way contrive to steal his thunder. If you came down to borrow a book – and he always had many library books on long loan – he would pretend to hunt for it on his shelves, always looking where he was sure the book was not, so that he could say, "It isn't here. You could have it if I could find it." If you knew just where to look for the book and pointed it out to him, he would grudgingly let you take it if you promised faithfully to return it promptly. This I know, for it happened to me times without number.

This fear of having his studies anticipated led him

into ways which have caused unlimited trouble to his followers. Specimens which he used for types of his new species were often so mangled that no further use could ever be made of them. Or else they were frequently put away, unmarked or incorrectly marked, so that it would be impossible to identify them later. If, by chance, the type was one which was correctly marked and placed in a jar in good condition, the jar itself would be very likely to turn up, years later, stowed away with totally unrelated species where the chances were a hundred to one it would never easily be found. My colleagues Henry B. Bigelow and William Schroeder have suffered sorely from these curious habits of Mr. Garman's.

He was a true pack rat of the first order. Everything that came into his room stayed. Scientific papers and books were often unpacked and put away on the shelves, but his desk drawers were filled with an accumulation of rubbish which it is hard to believe could exist. The crusts of his daily sandwiches were for years put into an enormous glass jar, perhaps to be fed to birds, but forgotten. The address labels from a weekly paper which for years unending he received through the mail were carefully cut off and hundreds upon hundreds stored away in a drawer. Envelopes, after the contents had been extracted, were carefully placed in another drawer. What use he thought could ever be made of these and the address labels is hard to imagine, but save them he certainly did, as he did all objects—old rubber shoes and the like.

His magnificent library he generously bequeathed to the M.C.Z., and when our librarian began to sort and catalogue it, she found many surprises. Innumerable items were old and rare. He had saved and accumulated quite a competency which was well invested, I suspect with Mr. Agassiz's advice, and he had searched through catalogues and second-hand book shops for years and picked up many publications which were of very great scientific interest.

Probably the reason for his habit of hiding things was his firm belief that other naturalists were all thieves. It was he who gave rise to the theory that Cope was the biggest thief in the world because he stole the whale. This animal, as I have told elsewhere, was washed ashore on Cape Cod, and by changing the label on a freight car to read Academy of Natural Sciences, Philadelphia, instead of Museum of Comparative Zoölogy, Cambridge, Cope got the whale. It is true that once there was vicious rivalry between scientists, and if anyone could beat Cope or Marsh in publishing a description first he felt it perfectly honorable to do so and, in fact, a triumph if he succeeded.

Why Agassiz took a strong liking to the queer fellow who met him, I have always heard, at the wharf in San Francisco in 1873, when the *Hassler* docked, and why he brought him back to the Museum and kept him on his staff, is difficult to understand. But he did just this, and Garman became a confidential member of Louis Agassiz's group of students and later an honored curator on Alexander Agassiz's staff. His early life and schooling remain a mystery to this day, but he was a natural-born linguist and a real scholar. Make no mistake on this point.

I am not going to attempt to catalogue the characters of all the various curators who haunted the Museum years ago. I will mention Dr. Charles Rochester Eastman, however, a polite, genial, and most erudite palaeontologist whose chief local claim to fame was the fact that he spent a long time in the Cambridge Street jail for having shot his brother. He finally was released on a plea of self-defense. It was a long time, however, before the release came.

Then there was Dr. Walter Faxon, a shy, old-fashioned scholar steeped in the classics, and a charming and versatile man. He was so absentminded that you might pass him on the stairway without a nod of recognition on his part and then walk into his room five minutes later with a guilty feeling that perhaps you had inadvertently hurt his feelings and find that he had never noticed that you had been anywhere near him. I often went to Dr. Faxon for information of one sort or another and was always overwhelmed with incredibility at the depth and breadth of his learning and his readiness to impart it.

Outram Bangs and Glover Allen were my very

warm personal friends. Concerning Outram I wrote the following just after he died:

This is a brown drab New England holiday afternoon in October. A few years ago Outram and Henry Bigelow and I would have been gunning together on this very day. I am quite alone, the Museum is deserted – how natural then that I fall to musing sadly of my friend's death, so recent that his very presence still permeates this my little back office, where daily we have had luncheon together since his first illness several years ago.

I had just written these few words, when by chance the opening lines of Bryant's *Thanatopsis* flashed into mind. I had not thought of this grand old poem for years. Who has? It is out of fashion now. "To him who in the love of Nature holds communion with her visible forms, she speaks a various language." How uncannily this fits just what Outram was! His interest in natural history was profoundly and singularly centered in the visible, as was his love of flowers and gardens and his skill in growing them, and his vast, I believe, unrivalled, recognition of the birds of the world. He seldom remembered their names, these he could always look up and well he knew where to find them; but he never forgot a bird, and he had seen about all of them between the museums of America and London and what once was "Tring" to ornithologists, and Berlin or Paris. The new forms unknown to him by autopsy he recognized either through remembering a description or through some curious intuition which he least of anyone could have explained. His instinctive analysis of a bird's relationships were equally sure, equally swift, and equally an expression of that inner knowledge of which he has left so meagre a record. Bangs would not generalize or put a surmise on paper, much less would he prepare formal scientific communications to read. He shrank fearfully from speaking before even a small group, even upon topics on which he could speak with prime authority. So also he shunned honors and recognition, and he was painfully harassed as he stood in Sanders Theater to receive the honorary degree which Harvard gave him years ago.

His influence will live, nevertheless, for he was a great instructor to the many young men who worked by his side and to whom he gave himself in full measure, pressed down and running over. The bird collection, nearly three hundred thousand carefully selected and perfectly arranged skins, is a tangible proof that this is true, and his labor has been appreciated by the host who have visited the Museum of Comparative Zoölogy to ask his advice or to bring their troublesome specimens to him.

His colleague, James Peters, has written of his early life, his collecting years, his love of the trout stream and the upland cover. We who have gunned with him remember him as one who shot with the greatest speed and accuracy, and who walked with a tireless stride that did in many a younger man. His splendid chest and narrow waist, through his whole life, bespoke the champion wrestler of his younger days. His weight had not changed for forty years.

I recall a day but a few years ago. It was in the Boston subway at the late rush hour. On the back platform of an electric car a poor Italian woman wedged up onto the step with a heavy bundle. The car was just starting when a flashily dressed youth of twenty years or more pushed her brutally to get on board. Bangs was next her, and like a flash his fist struck out and the man fell limp to the pavement as the car fast gathered momentum. 'I think that will teach him manners,' said Bangs, as he rubbed his fist, and then never gave the matter another thought, nor did he breathe the faster.

As for Glover, it is hard to believe that I will not find him there the next time that I go to his room. He was so retiring and modest that although I was in his quarters certainly several thousand times I never remember his coming to see me in my office except most cheerfully and politely when I called upon him for something which required a personal visit.

I remember when my friend Doctor Boschma, Director of the Leiden Museum, came here to represent his University at the Harvard Tercentenary, he spent several weeks at our house in Beverly Farms, commuting to Cambridge with me to work in the Museum. After luncheon one day in the "Eateria," at which a considerable proportion of the Museum staff were present, he said to me, "You Americans are very informal. You all call one another by your first names. That would never be done in a museum in Holland." I suppose this is true, for I remember once being roundly snubbed when making a call on a Dutch official in Java without first putting on a black coat. I have always thought that our Dutch colleagues see intellectually eye to eye with us Americans more completely than any other group of our European colleagues, but formal they certainly are.

The Museum of Comparative Zoölogy at Harvard College wears this dreary and meaningless style because the name was saddled on it concomitantly with a bequest of $50,000 received from the estate of Francis Calley Gray. It could get rid of the name only by returning the money to Mr. Gray's heirs, and this it cannot afford to do, insignificant as the sum appears today. In common with every department of Harvard University, the Museum has suffered bitterly from the strings which have been attached to the moneys bequeathed to it.

However, as I think it over, this handicap in the long run has really been a benefit to the Museum. No one has ever sought employment here for the sake of the salary. We have therefore a staff of servants who are wholeheartedly devoted to the Museum, and because of this community of interest it is the most pleasant and congenial place in which to do scientific work that can possibly be imagined. That this statement is not based on sheer imagination is proved by the myriad letters which I have received from workers in other institutions explaining that they would like to transfer their allegiance and join our staff.

It has been a well-established custom from the very beginning to work for the Museum without stipend if one could possibly afford to do so. I began once to look through the file of Annual Reports to see how many volunteer naturalists have worked here during the last eighty years or more. The number is unquestionably a large one, but the informality existing concerning the matter of appointments and titles was often so confusing that no complete roster can be made up at this late date, however interesting such a list would be from the point of view of museum history. In the forefront, of course, stand Alexander Agassiz himself and his gifted brother-in-law, Theodore Lyman, who not only aided him but, being a man of affairs as well as a naturalist, prevented the old professor, Louis Agassiz, from pushing himself and the Museum so deeply into debt that the institution might have been forced out of existence.

Mr. Lyman also mastered the taxonomy of the

group of the brittle stars, or Ophiuroidea, to a point where he became the recognized world authority. He was asked to write and he prepared the volume dealing with this group in the series of *Challenger* Reports. There could be no greater compliment than to have this request come from England. Mr. Lyman described more species of brittle stars than any other person who has ever worked with them, before or since the period of his activity.

Anyone who has ever looked at the drawings of the immature stages of dragonflies done by Louis Cabot will realize that here was another naturalist who was a first-class artist as well, and it is a pity that his interest was so short-lived. Dr. Walter Faxon, whom I have mentioned, was one of the mellowest and most widely learned savants with whom I have ever come in contact. One of the first carcinologists of his time, with memoirs on deep-sea crustacea and on the crayfishes to his credit, he was equally a first-class ornithologist. An erudite Latinist, he also made a collection of Shakespeariana which was a welcome increment to the University Library after his death.

Outram Bangs, of whom I have also spoken, came to this Museum originally to be a volunteer Curator of Mammals. He came after the period of his own most active collecting was finished, and since he found resources slender here and since Mr. Agassiz took little interest in any of the collections of land vertebrates, there were not many mammals here to be cared for beyond what he brought with him in his own excellent collections. Moreover, not many years passed before it became quite obvious that Dr. Glover Morrill Allen was going to be the outstanding mammalogist of the group here in Cambridge. So it was fortunate for the Museum that Bangs's interests had early turned toward ornithology. When he joined the staff, Mr. William Brewster, also of course a volunteer, had been Curator of Birds for many years, but Mr. Brewster had his own private museum adjoining his house on the corner of Brattle and Sparks streets. Here he labored for years preparing the *Birds of the Cambridge Region*, published as a Memoir of the Nuttall Ornithological Club in 1906. This is no ordinary local checklist. It is a monumental work, the finest of its kind which has ever been written. Brewster recast every sentence times without number, and his writing always combined accuracy of statement with great charm of style. His great and constantly growing collection he kept at home. It came to the Museum by his bequest upon his death, in 1919.

Bangs found at the Museum a motley collection of old wrecks, so far as the birds were concerned. To be sure, there were a good many nuggets in the sand. Many of Alexander Wilson's birds came ultimately to the Museum after various vicissitudes of which I have told elsewhere. These specimens have been identified with painstaking care and described by Dr. Walter Faxon in a Bulletin of our Museum. There were a few interesting birds collected by Samuel Garman when he visited Lake Titicaca in Peru with Mr. Alexander Agassiz in 1874. How he contrived to tear the skins off birds and have them hold together yet present such a frightful appearance is hard to understand. From among these specimens Dr. J. A. Allen described a few interesting forms. This he did back in the days before Bangs's advent, when, for a short time, he himself was the Curator of Birds in this Museum, before going to be curator in the Museum in New York.

There were other small suites of specimens which contained some very nice things. Early exchanges had been made with the Museums of Paris, Dresden, and Leiden, and a selection of birds from the Hume Collection had been received – I suspect, far fewer of these than Hume intended us to have when he told Mr. Agassiz that his birds were to be divided between the British Museum in London and our Museum here. Probably if a representative of our interests had been present when the division of the collection was made we would have received a good many more than we did. No such representative was sent, however, and our share of the Hume Collection was a sorry remnant.

Bangs, in spite of the fact that he was formally Curator of Mammals, promptly began resolutely to build up the collection of birds as soon as he came to Cambridge. He did this without much encouragement, too, for Mr. Henshaw, then Director of the Museum, was a stickler and wished each curator to stay put, so to speak, and not wander from his appointed field.

I remember once writing a paper describing some new birds which I myself collected on one of my journeys to Cuba – made at my own expense I may add – but I got such a thundergustuous tongue lashing for venturing away from the field of herpetology that I removed my name as author from the proofs and put Bangs's name in its place. My old friend, Mr. Charles F. Batchelder, who edited and still edits the Proceedings of the New England Zoölogical Club in which the paper was to appear, was surprised and still chuckles when he recalls receiving the proof, returned with the name of a new author at its head. Bangs's name appeared as author of the paper when it was published, and peace was restored, which was the thing most obviously desirable.

Bangs and I worked hand in hand to build up the collection, and in all the years of our intimate association I have never known a person who was more selfless and more completely devoted to the collection, which he gradually adopted as if it were his own. When Mr. Brewster died Bangs was made Curator of Birds and Allen assumed full charge of the mammal collection.

Building up a mammal collection is quite a different task from that of building up a collection of birds.

Many mammals are large, hence the securing and preparing of their skins is very expensive, and there have always been far fewer amateurs, hence fewer alumni of the University, who have been interested in seeing the mammal collection grow than have aided with the collection of birds. Mr. John Eliot Thayer was an outstanding exception, and during the many years when the veteran collector Wilmot W. Brown was working for him in the Southwest of the United States and Mexico, he collected mammals as well as birds. But for many reasons the collection of birds grew much faster, until today it is far larger and far more representative than is the collection of mammals. This is not to say that the collection of mammals is not a good one considering the money which we have had to spend on it.

I don't know whether it was Bangs or I who first began to have dealings with W. F. H. Rosenberg, once a commercial collector at Ybarra in Ecuador, who later moved to London and became a famous dealer in natural-history objects. Bangs and I both visited Rosenberg on several occasions. We established most pleasant relations with him, and for many, many years he gave us the first chance to purchase the finest and rarest specimens which fell into his hands.

We kept him supplied with carefully prepared lists, made as complete as we could make them, showing the species of birds, reptiles, and amphibians which were unrepresented in our Museum, and for years he gave us the refusal of the best that came his way. Thus we picked up many rare birds and not a few types that fell into his hands when private collections in England were dispersed. Types were what Bangs and I were constantly on the lookout for.

No story of the Agassiz museum, as I love to call it, is complete without the yarn that should be put on record against what may possibly happen in the distant future. Mr. Agassiz arranged with Professor Ward of Ward's Natural Science Establishment in Rochester, New York, to proceed to the Argentine and collect vertebrate fossils in the so-called Pampean fossil beds. The result of this enterprise may be seen in the Museum to this date in those splendid mounted South American fossil vertebrates. Professor H. F. Osborne of the American Museum spent a vast amount of energy trying to persuade Agassiz, then Henshaw, and then me, that these animals were so significant and important that they should be on display in the Museum of Natural History in New York where they would be seen by millions rather than in Cambridge where they would be seen only by thousands. Our giant armadillo Panochthus was the specimen he particularly coveted. Despite his pleas, this remarkable individual along with the Toxodon and the giant sloths still remains in Cambridge, where they have been on exhibition for over forty years.

What has been forgotten or almost forgotten is the fact that this collection represents but a small part

LOUIS AGASSIZ AND BENJAMIN PIERCE

PROFESSOR WILLIAM DANDRIDGE PECK

of Ward's booty. By far the larger part of his collections were placed aboard a sailing ship which left Argentina for New York and was wrecked off Fire Island. Wrecks of old sailing ships last for a long time under salt water, nevertheless sooner or later this wooden ship will surely disintegrate and it is more than just possible that someone in the future, dredging in the environs of the Hudson River Gorge off New York, will some day appear with a find of vertebrate fossils of strong South American affinities. I want to put this on record in case the event occurs within the life of the paper on which this is printed.

The material which we have on exhibition in Cambridge represents those findings which Ward came upon after the other shipment had been made and consists of the specimens which he himself brought back with him on his return from the expedition. Needless to say, I have spent many idle hours wondering what wonders may have been lost.

Among the innumerable Englishmen who from time to time have visited America offering lectures on all sorts of diverse topics was Alfred Russel Wallace. One of the results of his trip was the publication, in the *Fortnightly Review* (September and October 1887), of an article entitled "American Museums." Curiously enough, although he traveled widely from coast to coast, he chose the Museum of Comparative Zoölogy and the Peabody Museum of Archaeology and Ethnology, both at Harvard College, for extended description. This was more extended and lavish praise than the exhibits in either of the two institutions merited, admitting even that the point of view in 1887 was different from that of today. The great majority of our visitors at the present time are attracted to the building by the fame of the glass flowers and ramble about in a more or less haphazard fashion after they have visited the botanical section.

In the course of Wallace's meticulous descriptions of the individual rooms and their contents he quotes from one of Mr. Alexander Agassiz's annual reports, where he says, "The great defect of museums in general is the immense number of articles exhibited compared with the small space taken to explain what is shown. The visitor stands before a case which may be exquisitely arranged and the specimens carefully labeled, yet he does not know, and has no means of finding out, why that case is filled as it is; nothing tells him the purpose for which it is there. The use of general labels and a small number of specimens, properly selected to illustrate the labels, would go far towards making a museum intelligible, not only to the average visitor, but often to the professional naturalist. . . . The advantage, therefore, of comparatively small rooms, intended for a special purpose and for that purpose alone, will overcome at once the objections to be made to large halls where the visitor is lost in the maze of the cases, which, to him, seem placed without purpose and filled only for the sake of not leaving them empty." Surely Mr. Agassiz had the Museum in Paris in his mind.

This also leads me to believe something which I long suspected, namely, that Mr. Agassiz had plans for this Museum which, owing to the pressure of his interests in connection with the Calumet and Hecla Mining Company and the long periods of time when he was absent from the Museum on voyages of exploration, he never carried into effect. Certainly no such system of labeling was to be found here. For years before I was appointed Director of the Museum, when a specimen was placed on exhibition, there it was and there it was meant to stay. Keys to the exhibition cases, necessary if specimens were to be removed for purposes of study, were jealously guarded, and once a name was assigned to a specimen, rightly or wrongly, and the specimen was put on a shelf in an exhibition hall, there the object remained, even though the alcohol dried out and the specimen faded to a pallid white.

I never can forget the feeling of joy and satisfaction which I had in having innumerable duplicate case keys made and allowing any curator to remove objects from exhibition, to make changes in installation, and to change labels freely and without consulting me. I knew the staff and I knew that if it was sufficiently competent to pursue research in connection with the various study collections of animals its members were certainly competent to handle the infinitely less valuable material which had been set forth for the edification of the public.

While vast numbers of descriptive labels have been laboriously prepared and the number of specimens remaining on exhibition is much reduced, there are, I am sorry to say, still a number of individual animals badly mounted and faded for which a single label gives only the Latin name and the indication of a locality. I have discussed this label problem elsewhere at some length.*

In 1927, President Lowell called me to his office and told me that he wished me to assume charge of the Museum on November 1.

One condition I determined at once to change, and did change at the earliest possible date. Our building was lighted throughout by open fantail gas burners. Since the house is not by any means fireproof in construction, it has been entirely thanks to the great care exercised by the staff that no disastrous fire has ever occurred. To be sure the danger was lessened by the fact that the Corporation, on the advice of the insurance underwriters, had had automatic sprinklers installed throughout the entire building and smoking was, and still is, forbidden. Nevertheless I breathed a sigh of relief when the last gas pipes were removed and a modern and thoroughly well-insulated system of electricity was installed.

* See "Thinking Out Loud," below.

There was an elevator shaft, and this was not enclosed, which was fortunate, for masses of worthless and undesirable material, found in several tightly locked rooms which we came to call the Glory Holes, were conveniently dumped down the elevator shaft until, at one time, the pile reached well above the second story of the building. This material was hauled out from the bottom of the shaft and sent to the city dump. Many of the discarded animals were simply placed on the lawn, whence they were carried away by boys and girls from those squalid sections of North Cambridge and Somerville which are far too near the Museum for the safety of our windows, of specimens which can be touched, and of labels which can be reached and torn loose from our walls. The elevator shaft was later closed in and a modern electric lift was installed, to the great comfort of all the staff, especially the more elderly members thereof.

At this time, by great good fortune, plate glass was relatively inexpensive, and since much of the original glass built into the exhibition cases was brittle and of very poor quality indeed, plate glass has made an enormous change in the appearance of many of the exhibition halls. Of course, all these changes were made in the days of the so-called Era of Inflation, more accurately termed the Time of Prosperity, in which fortunately I shared with many others. Had my opportunity come ten years later the Museum would be a totally different looking place from what it is today, although I'll confess that it is not yet ideal from a thoroughly modern point of view.

Early in my incumbency I became dissatisfied with the quality of a number of casts, principally of fossil animals, which had accumulated and been placed on exhibition with the passing of the years. Indeed into many of these too much imagination had entered; they were based on fossil material much less perfect than that which has been discovered in recent years, and the restoration was faulty. I came to the conclusion that, considering our limited space for public exhibits, we would show nothing preserved in alcohol and no casts, with, of course, the mental reservation that if there were good reasons to be inconsistent, to break the rules would be good fun. As a matter of fact, to the no-cast rule I have stuck pretty firmly. I could not refrain, however, from putting on exhibit last winter a couple of the beautiful replicas of sharks, prepared by Pflueger of Miami. They are absolutely perfect in every detail and they have decorated the Fish Hall mightily; and good mounted sharks are difficult to make and hard to procure.

Our Museum differs from all others in that the exhibits, most of them, fall deliberately into two distinct categories: those arranged to show the animal life of the several zoogeographical regions as they are commonly recognized, and those arranged in synoptic collections representing several different groups of animals. Originally most of the exhibition halls had a narrow gallery with a guard rail which was far too low, and bad cross lights. Light came in from the windows of the floor below as well as the windows in the gallery itself, and moreover these galleries were inaccessible to the public, hence practically unvisited. The material on display represented material from Africa, South America, Australia, and so on – groups of animals in which the public was not in the least interested. These exhibits were preserved mostly in jars of alcohol, and most of the contents were specimens faded beyond recognition. What we really needed very badly was additional space for our magnificent and rapidly expanding study collections and for the greatly increased number of visiting investigators who were attracted to the Museum by various changes in policy. We therefore determined to floor over all these galleries, and thus ten additional large rooms for research work were gained.

One individual specimen almost blocked the scheme. Once when Mr. Agassiz was in London he visited Roland Ward's Studio and purchased a beautifully mounted specimen which I verily believe was the father of all giraffes. I have seen many living giraffes, myself, in Africa, but I don't think any one of them approached our problem in size. I decided where it was to be relocated, which was to be out in the great main hall where the three big and excellent whale skeletons hang from the ceiling. I planned to retain this gallery for a synoptic collection of birds. To make a long story short, we tipped the giraffe over, laid it on its back, and contrived a sling of heavy sailcloth. With a gang of men, struggling and staggering and pushing and pulling, we moved the giraffe, with a clearance of less than half an inch between the sides of his belly and the jambs of the doors through which he had to pass. He is much more of a spectacle in his new location, where he may be seen from a distance and from all sides, and I always notice the awe-struck delight with which the classes of little mites from the Cambridge public schools gaze up at his vast bulk when they are taken by Miss Crawford, of the Children's Museum, for their museum walks.

Incidentally, these museum walks are one of the reasons which have stayed my hand time and again from advocating a further contraction of the space devoted to public exhibits. The rapture and enchantment which is obvious when Miss Crawford takes these groups on a tour of the building is sufficient alone to justify maintaining the public rooms of the Museum. This is a great and little-known public service which the University provides for the school children of Cambridge.

The general systematic halls of reptiles and amphibians were the most unsatisfactory in the Museum. The several thousands of faded alcoholic specimens on the shelves were soon relegated to the study laboratories, sorted, and either discarded or, where the specimens were worth it, added to the research collection. The same was done with the fishes. From various sources we began to assemble new mounted material for the rooms, which are not wholly satisfactory but are much better than they were.

We determined to set up a hall adjoining the reptile hall as an Alexander Agassiz Memorial. In this we placed the beautiful models of the coral islands. Unfortunately, these are vastly larger than they needed to be, but they are exquisitely made. Then, in the wall cases surrounding these, we set up representative collections of brilliantly colored fish and invertebrates such as are found about coral reefs. Consequently, this is now quite a gay and attractive hall.

Adjacent to the hall of fishes we installed a synoptic hall of invertebrates, showing a few choice members of each group. Microscopic protozoa, naturally, are shown by models; and in some of the cases drawings and colored illustrations are used when the original actual material is impossible to prepare for exhibition. Doctor Elisabeth Deichmann, Doctor Bigelow, Professor Nathan Banks, and Mr. W. J. Clench did yeoman service in setting up this most instructive but not at all spectacular or popular exhibit. It is visited and found useful, however, by the relatively advanced classes in biology of various institutions of this neighborhood, year after year. The synoptic collection of insects is really noteworthy, in particular because almost every family of the whole enormous order is represented by a few carefully chosen specimens.

All of this naturally took years to accomplish and, as is inevitable with the passing of time, one's interest wanes. My colleagues and I have been woefully negligent about sandpapering, from time to time, the exhibits which we set up. As is natural, every one of us prefers to engage in research work, and the amount of this which has been turned out during the last two decades has been extremely gratifying and a source of great pride. I may also add here that if it had not been for the perennial financial assistance of Mr. George Russell Agassiz it would have been quite impossible to publish the great volume of manuscripts offered. To this fact I must add my gratitude to my colleague, Ludlow Griscom, for assistance in directions where my competence is not outstanding, such as the budget, which for years has had to be made at a time when I was occupied in the tropics, and editorial work, wherein I find myself more competent in picking flaws than in constructive criticisms.

CHAPTER III

The Natural History Rooms in Boston

THE Boston Society of Natural History held its first meeting at the house of Dr. Walter Channing on the 9th of February, 1830. At this meeting a nominating committee was appointed, and Thomas Nuttall was selected to be the first president of the Society. Prior to this date, Boston had shown little interest in any of the several branches of natural history, nor indeed had Harvard College been as foresighted as Yale College, for instance, in developing teaching and more especially research in such matters as geology, botany, and zoölogy. Of course it is not literally true that there had been no activity at all, but it was puny by any standard. The new Society was to exert an influence much more significant than any of its members would have thought possible.

I have always thought that one of the most interesting possessions of the Society was a letter from Charles Lyell to Mr. George Ticknor, which read as follows:

> I am trying to negotiate with Mr. Lowell for a course of lectures from the celebrated ornithologist and Swiss naturalist and writer on glaciers for

> 1845–46, but perhaps all are filled up. Charles Buonaparte, Prince of Canino, has offered to take him to the United States, as he visits it with his son this year. I am sure Mr. Lowell will do it if he can, as I have answered for his English being passable. You will be much pleased with Agassiz, and his visit will be most useful, as it always is to us when he comes here. The British Association has thrice voted him sums of money to describe our fossils.

There is no question that this short line accounts for Louis Agassiz's coming to America and for all that he contributed to the scientific life of the country as well as to educational methods, to say nothing of the excellent scientific work which he carried on after he came here.

In 1847 Dr. Samuel Cabot reported that a suitable edifice for the Society was available, inasmuch as the Massachusetts Medical School was for sale, and this building became and remained the home of the Society until it entered the building which it occupies today. This event took place after that session of the Great and General Court of Massachusetts which in the winter of 1860–61 granted the site of the present building to the Society.

In February 1930 the Society became one hundred years old. In commemoration of the event a little book entitled *Milestones* was printed, edited by Percy R. Creed, the cost of the edition being borne by a small group of members. This was beautifully printed by the late Mr. Updike of the Merrymount Press, and it reproduced a number of excellent portraits owned by the Society, of such men as Thomas Nuttall, John Collins Warren, Jeffries Wyman, Louis Agassiz, and many other worthies of former days. It also illustrated a few of the highlights of the collection, such as a pair of American oyster catchers that were shot and presented by Daniel Webster, and some birds' eggs that were found and given by Henry D. Thoreau.

This little book recites a story of growth and decline. It tells about the celebration when the original portrait of Alexander von Humboldt, painted in Berlin in 1852, was presented by the Reverend Robert C. Waterston at the time of the Humboldt centennial, September 14, 1869. A poem on this occasion was contributed by Oliver Wendell Holmes, the last stanza of which read:

Bring the white blossoms of the waning year,
Heap with full hands the peaceful conqueror's shrine
Whose bloodless triumphs cost no sufferer's tear!
Hero of knowledge, be our tribute thine!

Agassiz was present and made an eloquent plea for higher education. He spoke as one who himself had undergone "the necessities which only the destitute student knows," for as a young man at twenty-four years of age he had, in Paris, come to the end of his resources and had been befriended by Humboldt, who remained his friend as long as he lived.

Many of the occupants of the presidential chair, who through the passing years guided the destiny of the Society, were distinguished citizens of the Commonwealth, and for years the institution was an active center of research, its meetings frequent and well attended. Little by little, however, the Society's activities were overshadowed by the growth of the Museum of Comparative Zoölogy at Harvard College. Both institutions have always been impecunious, and both institutions have operated with a capital endowment about equal to the annual income enjoyed by the great metropolitan institutions elsewhere. But the presence of more virile leadership in Cambridge, and the fact that Harvard College naturally attracted a great number of distinguished scientists, caused the University Museum in Cambridge to grow at the expense of the museum in Boston, so that the latter institution had to cut its cloth according to its waning resources. It is a strange and curious fact that the Bostonian is generous beyond measure to music, art, and charitable institutions of all sorts, but his interest in natural history has been extraordinarily conspicuous by its absence.

It is hard to believe today that until 1907 the Society received a small income from admission fees to the so-called exhibition halls, although for many years they have been open to the public two days a week.

I think that the decline in the fortunes of the Society may be said to have begun with the death of Professor Alpheus Hyatt, long its Curator, which took place in 1902, for it was about this time that it became clear that the fields of the museums in Boston and Cambridge were being cultivated in far too much the same way. This should never have been allowed, and should have been easy to avoid.

Naturalists, however, are queer, temperamental people. For example, when I first came to the Harvard University Museum, we housed, besides the Museum, the zoölogical departments of the University. There was the most conspicuous possible lack of coöperation, to make a monumental understatement, between these two groups of people, the Museum staff and the University faculty. This was in part, at least, because Mr. Alexander Agassiz and Professor E. L. Mark did not exactly admire each other — another acme of understatement. Professor Mark was always a dévoté of the latest zoölogical specialty fashionable at the moment. Spermatogenesis and ovigenesis happened to be in the ascendent just at the time of which I write. I was not highly thought of, because I wished to become a taxonomist and worked in the Museum and wrote a thesis on the distribution of animals in the East Indies. My examinations were passed largely owing to the kindness of Leonhard Stejneger and Carl H. Eigenmann, who helped me with their expressed approval.

Now, if this sort of situation can exist within a single institution, do I need to expound to you what may happen between separate and distinct institutions? This, in part, explains why Harvard University and the Boston Society did not always have very intimate associations. The Curator of the Society, Dr. Alpheus Hyatt, a most excellent naturalist, for many years longed for a much more intimate relationship with Harvard in the form either of a professorship or a position in the University Museum, which he never got, for in Cambridge he was not *persona grata* with either group. Childish and stupid you may say, and my answer would be inclined to be the old German saying, "*Traurig nichtdestoweniger wahr.*" Hyatt died, C. W. Johnson from Philadelphia was called as Curator; and finally he died.

When I joined the Boston Society of Natural History in November, 1902, the museum, contained in what was then called the Natural History Rooms, was in very truth the last word in what apparently was a studied effort to demonstrate what a museum should not be. The exhibits were so dingy, so overcrowded, and, I may add, so revolting, that it was widely rumored that recalcitrant children of the families living in the Back Bay were dragged in and walked through the Natural History Rooms to strike terror to their hearts as a bitter and long-to-be-remembered punishment. I was not squeamish in those days and had certainly much more curiosity than the average,

UPPER: RECENT ACTIVITY IN THE BOSTON MUSEUM

LOWER: UNSUITABILITY OF ARRANGEMENTS IN THE SAME MUSEUM

UPPER: THE TYPE OF CAROLINOCHELYS WILSONI AS IT CAME FROM CHARLESTON

LOWER: RENOVATED EXHIBITION HALLS IN THE MUSEUM IN BOSTON

but the injected, desiccated, and varnished human dissections, part of the highly prized Jeffries Wyman Collection, to say nothing of the human fetuses and monsters, were so utterly repellent and revolting that after looking at them I had a creepy, half-guilty feeling, such as I had as a boy when I sneaked off and visited the morgue in Paris. Those horribly stuffed primates, leaning drunkenly against upright bits of trunks in postures which they never assumed in life but were made to assume when mounted because in an upright position they took up less room, are vivid in my memory to this day. I may have been a little more sensitive about these matters than other people. Apparently Glover Allen, who asked me to join the Society and who was then its secretary, and Dr. Robert T. Jackson, my adviser as a freshman and long a member of the Society, as were his father and many of his relations, did not feel quite as strongly as I did about the unsuitability of all this material for public exhibition. However, no hodgepodge I have ever seen, in Paris or elsewhere, has ever demonstrated more clearly and certainly what a museum should not be. I early determined to make a sincere effort to change all this, and I had a competent ally in the person of my colleague Dr. Henry B. Bigelow. At the time of which I am writing he was a graduate student working for his doctor's degree, whereas I was a little younger and still an undergraduate.

Dr. Charles S. Minot, who was president of the Society, was a courteous, charming gentleman and a sympathetic and willing listener. It was in 1913, near the close of his presidency, that I broached the idea of limiting the exhibits and the research activities of the Society to the New England field. Even as early as 1914 it was becoming obvious that the investments of the Boston Society, which were almost entirely in railroad securities, were not as safe as it had been hoped they would prove, and moreover it was becoming more and more obvious that Boston lacked interest in its Museum of Natural History. This fact which I have already mentioned, is one worth dwelling upon, since it is very difficult of explanation. The gifts which Bostonians make to music, art, and charity taken all together are the largest gifts per capita of any city in the United States. Many smaller centers, however, such as Buffalo, Milwaukee, and Newark, to mention only a few examples, are much more generous in the support of their Natural History Museums than is Boston. Memberships have always been difficult to secure and the turnover in the form of resignations and the necessary search for replacements has been most discouraging. These facts all played their part in suggesting a more restricted program. There was certainly no object in the attempt which was being made to do exactly the same sort of work which was being done in the zoölogical and botanical institutions connected with Harvard University, three miles away, where there were an infinitely more suitable building, a more adequate endowment, and a far larger staff, to say nothing of research material which in amount had grown to be far greater in variety and scope than any of the collections in Boston.

Needless to say, the policy of restriction was vigorously opposed, and there were many who believed then, and there may be some who believe now, that the proposal represented a case of studied determination on my part to ruin one museum for the benefit of another. Since it was quite obvious that there would be less interesting New England material which could be sent from Cambridge to Boston than there was in the way of exotic collections of natural history to be sent from Boston to Cambridge, what probably helped more in turning the trick than any other single factor was Henry Bigelow's announcement at the Council meeting that the members of his family had no objection whatever to the transfer of the Lafresnaye collection from Boston to the Museum of Comparative Zoölogy.

This collection, formed by Baron Lafresnaye, an eminent French ornithologist, was purchased by Dr. Henry Bryant and given by him to the Society in 1866. It contained the type specimens of over 700 species from the American tropics, most of them described by Baron Lafresnaye. Dr. Henry Bryant was Dr. Bigelow's grandfather, and his private collection of birds had already been turned over to the Agassiz Museum.

Dr. Bigelow became a member of the Council of the Society in 1911. I became a member of the Council in 1913, and in the Museum and Library Bulletin appearing in June, 1914, the Curator, Mr. Charles W. Johnson, reported as follows:

> The increasing need of more space for fully displaying the New England collections has finally resulted in the decision of the Council to transfer to the Museum of Comparative Zoölogy such non-New England specimens as are not needed in the Society's present scheme of making the local fauna, flora, and geology the chief feature of its exhibition and study collections. The Cambridge Museum in its turn will reciprocate by giving to the Society additional New England specimens, special material for the synoptic collection, and assistance in developing certain New England groups.

Already the Lafresnaye types of birds, together with a number of other specimens of unmounted birds from this collection, the birds from the old Peale Museum, those of the Wilkes Exploring Expedition, some 3,000 miscellaneous bird skins, and a number of European fossils had thus been given to the Museum of Comparative Zoölogy. Shortly after the Lafresnaye birds were moved to Cambridge, Bangs associated with his friend Thomas Edward Penard in a project to write up the story of Lafresnaye and the collection with a view to publication. What became of the biographical section of the manuscript was for years a mystery; in 1944 Mr. J. L. Peters found it. It will be published now in due season.

Tom Penard, as we all called him, was one of the most painstaking scholars who have ever worked here. It is a pity that his profession, for he was an electrical engineer, prevented him from devoting all his time to ornithology. Born at Paramaribo in Dutch Guiana, he knew the birds of his homeland at first hand, and he left a fine collection to the Museum at his death. He was an excellent linguist and had a most extraordinary memory. Bangs was extremely fortunate to have so well-rounded a scholar to work on the history of these Lafresnaye birds with him. It is a great pity that his death delayed the appearance of the study to which they both devoted so much time.

I have a mass of material left at Penard's death concerning bibliographical details, such as the matter of the exact dates of the publication of Lafresnaye's names. The final decision concerning the status of each one of these type specimens under discussion was made by Bangs, and the details concerning each decision are embodied in his paper on the "Types of Birds now in the Museum of Comparative Zoölogy," published in the Bulletin of the Museum for 1930. The amount of skillful checking and sleuthing out of the evidence in each individual case was a monument to the painstaking care with which Bangs and his friend searched the available literature as well as all sorts of contemporaneous sources of information. Thus the last word has probably been said concerning this matter which has given rise to so much discussion. Since the collection from a scientific point of view was one of the most valuable ever brought to America, we owe Bangs a great debt of gratitude for the extraordinarily careful work which he did in this connection.

I quote from Bangs, first from his Bulletin of 1930 and then from a draft of an introduction to a paper concerning the types of Lafresnaye, the manuscript of which I found in the Museum.

> The types of a few of the species described by Lafresnaye, which apparently should be in the collection, cannot now be found. This, however, is not surprising, as types were not valued formerly as they are today, and if poor specimens, or damaged from one cause or another, were very likely to be discarded and replaced.
>
> Lafresnaye was a much better systematic ornithologist than the record of the synonyms made by him would indicate. If one glances at the names in the following list one will see that more than half of those that got into synonymy do so by months or possibly by a year or two only.
>
> In Lafresnaye's time the greatest source of new

birds was the enormous supply of "trade skins" constantly pouring into the hands of the Paris dealers. While what we call luck seemed always against him, it must be remembered that Lafresnaye lived in the country, in those days a real journey away from Paris, and, therefore, he was often just a little later than someone else in securing some new bird. Also, I fancy, published descriptions were slow in reaching him. Several times I have read a complaint to that effect written by him on a label.

It has always been supposed that all of the types and cotypes of species described by Lafresnaye from the Delattre collection which was bought by Dr. Wilson and presented to the Academy of Sciences of Philadelphia were in the collection of that institution. Labels for certain specimens in the Lafresnaye Collection clearly show that Lafresnaye, by some arrangement, retained some of the Delattre birds for himself. In the following list we claim types or cotypes of several of the species. In every instance where we have done so, the evidence has been submitted to Dr. Witmer Stone [of the Philadelphia Academy] and we claim the type or cotype as it may be, with his full approval.

In 1925 I became President, and remained so until I was made Director of the Museum of Comparative Zoölogy at Harvard College, when Mr. Lowell felt that I should devote all my efforts to the situation in Cambridge. I resigned, but was constrained to resume the presidency in 1940, and continue to hold this office pending the finding of a more suitable candidate.

When Henry Bigelow and I, as youngsters, managed to get elected to the Council of the Society we were faced with an old building, fearfully overcrowded, and an organization which for years had laid stress upon supporting research work. The available funds would not allow this to be done and decent exhibition collections to be maintained as well. It was the prated fact that "the prestige of the Society abroad must be maintained and papers based on research published." The Museum down to 1919 accepted material of any and every sort. This policy brought about the situation which I have outlined. What happened afterwards may now be told, and the history of recent years leads one to believe that today the Society is setting forth along a new road of increasing usefulness.

It is a pity, as a matter of fact, that this transfer of the Lefresnaye collection from the Museum in Boston to Cambridge did not take place many years earlier. Some of the Lafresnaye birds had suffered terribly from the fact that the whole collection was stored in cases which were in bright light and which were not too tight and dustproof. The most important matter connected with the removal was the fact that Outram Bangs and his friend Thomas Edward Penard were constrained for various reasons to study each bird in the collection critically. A large number were discarded as worthless. Lack of space in our Museum in Cambridge has forced us to be somewhat choosy, so to speak, and to eliminate undesirable specimens more ruthlessly than is the policy in other museums with greater available storage space. Bangs and Penard found that there were exactly 350 types in the collection of thirteen different ornithologists, no less than 269 being by Lafresnaye himself.

Thus after eleven years the first stage of what I had hoped to do with the Boston Museum was accomplished. Clearing out and moving the collections of foreign material made room so that some rearrangement was possible. Some rooms were completely emptied so that new cases might be built, and then the material was moved back again and the process repeated. With the help of Dr. John C. Phillips and other friends, new and improved installations began to appear. Today the Museum is as attractive as it can possibly be expected to be. Clean and relatively well lighted, it has drawn increased public attendance. Extension activities of various sorts are highly successful and extremely promising. We hope that they may have the result of inducing additional support.

The Boston Society of Natural History has lived for long years from one day to the next in hopes of a new building. It occupies, at the present time, as it has for years, a flimsy, ill-built, structure which is, I believe, without question the one most ill planned and generally unsuited for museum uses of any which is occupied by such an institution anywhere in the world. There are reasons to suppose that possibly something better may be forthcoming but I don't think there's very much to be gained by counting chickens before they are hatched.

This institution is not now, nor can it ever hope or expect to be, a research museum or a museum with a staff of specialists dedicated to accumulating collections upon which investigations will be based. The Harvard University Museum, on the other hand, is staffed by experts, and with its enormous collections, world-wide in scope, is conditioned to be a research institution. These two museums should try hard to complement each other intelligently and not to compete. The only conceivable future situation which might entirely change this picture would be the ultimate shrinkage and death of the privately endowed institutions and a possible increment of governmental and state aid which might ultimately cause the Museum at Cambridge to fall into the possession of the Museum at Boston. This is a possibility which seems to me so remote that it is hardly worth while devoting much attention to at present. I prefer to visualize the Society as continuing, in the future, to do just what it is doing now, only more attractively and with better equipment. Personally I do not anticipate that it will ever have a very large endowment. Nor do I believe that Boston will ever suddenly become more "natural-history conscious" in the future than it has been in the past.

The Museum in Boston is, I believe, destined to be a museum devoted to public entertainment and instruction, especially of young people, *a glorified and most honorable adjunct to the public school system*, as well as a center for popular adult education in science. This, I believe, should be the avowed destiny of the Boston Society, and an aim for which, before all other, it should constantly strive. Nor do I believe that it should ever attempt to maintain in the future anything more in the way of a library than just what is needed to furnish standard books of reference and other books of a general nature to the entertainment of its members. There is already obviously far too much library duplication in the greater Boston area, and the sale of the foreign serials which was made a year or so ago was surely a step in the right direction and one which might very well be duplicated in the near future. A large part of the library was long duplicated by many other scientific libraries in the immediate vicinity, and the disposal of some of its books to a far western University has been to advantage in that stackage has been freed and badly needed space thus vacated to care for an increase, not in the scientific serials, but in the books of popular reference read by our members who drop in from time to time to browse among the shelves. The library has been fitted with comfortable armchairs, and since the windows are enormous, the room is light, and is, indeed, one of the pleasantest reading rooms anywhere in the Greater Boston region.

Changes gradually come if one but has patience, and in Boston the face of nature has changed. While our sparkling young Director, Bradford Washburn, is absent in the service of his country, Miss Margaret Baker has filled his place with industry, self-sacrifice, and intelligence. The following account of the dawn of a new day comes largely from a report to the trustees made in 1945 by Miss Baker.

The Museum Junior Associates, an organization which was started last fall, is a club for boys and girls who come regularly to the Museum on Saturdays. It boasts to date 250 members. Saturday is young peoples' day at the Museum. Through the Main Hall and among the exhibits all over the building you will find youngsters crowded around exhibit cases filling out sheets of questions and making drawings of the animals. The Junior Explorers, a club for grammar school children, is in session Saturdays from 10 in the mornings until closing time at 5. They work in groups or singly on the subjects of natural history which interest them most. Club members often give lectures, make collections, construct small habitat groups, and during the good weather go on field trips. Live animals are kept at the Museum only for a few weeks before being released, so the children can observe their behavior and learn something about their care.

The Explorers Club is for High School boys and girls. In this club, as in the Junior Explorers, a sponsor acts as adviser, but the club is run by the members themselves. Field trips are one of the most popular activities, but during the winter months lectures by the members constitute a large part of the program. This group has recently become affiliated with the Science Clubs of America, and through the bulletins of that organization has become interested in experimental work in genetics.

Weekday clubs have also become popular at the Museum. Groups on painting, animal puppets, and general nature study have attracted many of the children living near enough to the Museum to come to the building after school hours.

During the three school vacations, in December, February, and April, special programs are planned during the day for the many children who come in at that time. Movies and nature games played in the lecture hall are particularly popular. School groups now come into the Museum often, and many more will come when transportation eases up. Before the war we had as many as six or eight groups a week, while now we run on an average of one or two a week, including adult classes and those from Teachers' College.

However, it is not only to the young people and their instructors that the Society offers special programs and active participation. One of the rooms downstairs has been recently rearranged as a place in which to hold temporary exhibitions.

Lectures also are part of the program for adults, but the most interesting, and Miss Baker believes, one of the most important, of the activities for adults is the Museum Workshop, started in October 1944 and held this year on Wednesday evenings. These classes are not lectures, but informal group meetings for laboratory work, and are designed for anyone who wishes to take up or pursue a study in one of the fields of natural science. Microscopes, books, glasses, tools, and our fine study collections of New England material are at the disposal of the students. The students work in the rooms that house the collections, so they can refer to a specimen at any time, or remove whole trays of bird skins, cone shells, or tourmalines, etc., to make a comparative study.

So to the child, fascinated by the animals and birds, on to the High School student leaning toward investigation, leadership, and club activities, to the instructor needing help in presenting the study of nature to her own students, and to the adult wanting to pursue a hobby or to learn more about the world in which he lives, the Museum holds out a hand of friendship and welcome, and around the desires, needs, and interest of these people it builds up and develops its new activities.

I've been a little disturbed over some agitation which is developing at the present time. This concerns the connection of the Boston Museum with mineralogy. Mineralogy is not a subject which can be made of interest to the visiting public except by means of what is necessarily only a small collection of especially beautiful objects, New England being what it is as regards its geological make up. This, with almost magical efficiency, the present staff has almost completed. A room has been arranged showing phosphorescence, displaying the gem stones from Maine of which the Museum possesses a unique series, and other objects which are attractive because of inherent loveliness. Otherwise, the general run of this material has no popular appeal and cannot possibly be made to appear attractive to the casual observer. Moreover, this has been proved over and over again. The mineral halls, in the museums which have them, are empty halls.

For years all the exhibition space on the lower floor of the Museum, where the library is also located, was devoted to an exhibition of geology and mineralogy. This is the best and most desirable exhibition space in the building, for the reason that it is most accessible to the visitor as he first enters from the street. Thus the best exhibition space long sheltered the material which has the least popular appeal of any collections possessed by the Society. This has in part been remedied by the installation of which I have just spoken, and one hall has now been set aside for temporary displays as they become available. This practice has been most successful, though for years the change was sturdily opposed.

The mineralogical resources of the Museum may be divided into two categories: there is the little assemblage of rare and unusual New England material which, of course, will always certainly be treasured and adequately cared for, for it has real scientific significance and intrinsic value and beauty; and then there is a great mass of run-of-the-mill minerals – specimens which can easily be replaced. This can be used to hand about to youngsters in teaching mineralogy, a subject which they greatly enjoy. If these specimens are spoiled or marred they are easily replaced. There is no special need of making an exhibition of them.

The geological collections, on the other hand, consisting of samples of building stones, paving stones, and the like, which admittedly have some value because they illustrate names applied years ago in economic geology and which standardized and typified the names bestowed, have at last been generally discarded. They are just about as interesting to look upon as the cobblestones in a city street, yet there were those who loved these specimens, and threats to get rid of them met with sharp and competent opposition. Mind you, too, all this material is duplicated with far larger and more valuable collections of just the same sort in the University Museum at Harvard, about three miles away as the crow flies. No one, however, ever cares or thinks much about this.

The Boston Mineral Club, an organization consisting of amateurs of mineralogy located in various educational institutions about Boston, has cast its protecting eye over this material in the possession of the Natural History Museum. So far as I can see, the members certainly have no very clear idea of what they feel we should do with it, except keep it just as it is.

I am frank to say that, as President of the Society, I, too, have no very clear idea of what the ideal destiny of the mineral collection may be. I don't think it can safely be stored away in any upper part of the building without its being very likely to cause the galleries to come crashing down, for all the upper parts of the structure in which we now carry on are in extremely poor condition; indeed, they have been flimsy from the original date of erection. I certainly do not want to have the collection continue to occupy the most desirable exhibition space we have. This should be devoted to a fine display of something such as the New England birds and mammals of which we have such beautiful exhibition material. Some of the Trustees are opposed to selling any part of the collection. One thing only I can suggest: that tiny chips from the great, bulky samples of building stones, etc., might serve the documentary purpose they illustrate just as well as the big blocks we have now. This would certainly help, but I'll venture a wager that if, and when, this contraction of volume is proposed it will promptly be opposed by someone. That has been the history of all change and attempted improvement in the life of the Boston Museum. I hope for constructive criticisms and suggestions, but nothing is more rare in this world than just this sort of assistance. I suppose perhaps the whole matter will have to lie over, just as it is now, until a new building is built, intelligently planned along modern lines, and strong enough to have an attic where all the unwanted material can be put away and left unseen and unheeded forever and ever Amen.

CHAPTER IV

Of Old Museums, Especially that in Salem

THE first natural history museum in North America was made up of collections in the possession of Harvard College. To be sure, this was a heterogenous assemblage of objects kept in a room in Harvard Hall and called the Repository of Natural Curiosities. Everything burned in 1764.

In 1769, however, a "Museum" was again established and assembled to replace the objects burned. This, growing from time to time and moved on a number of occasions to different buildings, finally was added to the material which Louis Agassiz began to assemble as soon as he came to America in 1847, and thus in reality was the beginning of the present Museum of Comparative Zoölogy.

Far different was the story of the origin of the Charleston Museum, long and erroneously thought to be the oldest museum in the United States. This it certainly is not. It is, however, the oldest institution which was founded with a definite program and inaugurated to serve a specific purpose. My friend Miss Josephine Pinckney of Charleston, whose great-grandfather was one of the first curators of the institution, helped me to find the statement recording what might be called the official birthday of the museum idea in America. On January 12, 1773, a meeting was held in the rooms of the Library Society in Charlestown where the following statement concerning the purposes of the Museum was set forth.

> Taking into their Consideration, the many Advantages and great Credit that would result to this Province, from a full and accurate NATURAL HISTORY of the same, and being desirous to promote so useful a Design, [the Society] have appointed a Committee of their Number to collect and prepare Materials for that Purpose.
>
> That this may be done in the most complete and extensive Manner, *they do now invite* every Gentlemen who wishes well to the Undertaking, especially those who reside in the Country, to co-operate with them in the Advancement of this Plan. . . . For this Purpose, the Society would Request such Gentlemen to procure and send to them, all the natural Productions, either Animal, Vegetable, or Mineral, that can be had in their several Bounds, with Accounts of the various Soils, Rivers, Waters, Springs, &c. and the most remarkable Appearances of the different Parts of the Country.
>
> Of the Animal Tribe, they would wish to have every Species, whether Terrestrial or Aquatick, viz. Quadrupedes, Birds, Fishes, Reptiles, Insects,

world. The contacts took place at a time when the artistry and artisanship of all native peoples everywhere was as yet uninfluenced by contact with the white man.

An early rivalry sprang up between the members of the Society, who vied with one another in the amount and variety of material which they brought home to Salem, and thus led to the magnificent and completely irreplaceable collections representing the handiwork of the various peoples visited. Larger and more varied collections exist, but there is no collection where the value of the material, object for object, is greater.

There were, of course, regions in the South Pacific where primitive man had little to offer which the Salem skippers wanted, where the people long remained savage and difficult to deal with, and where the regions were particularly unhealthy. For this reason many parts of New Guinea, New Britain, New Ireland, the Solomon Islands, and the New Hebrides have remained in such condition that recent expeditions have brought back fine and varied collections. They have indeed been able to secure material as well made when it was picked up say twenty years ago as it would have been had the collections been made one hundred years or more ago. Consider as an example the collections made not so many years ago in the Admiralty Islands for the Field Museum in Chicago.

How different the situation with the Polynesian Islands where the natives were immediately friendly to the white men, where missionaries spread far and wide at an early date, and where the conditions under which the natives lived immediately began to change! The Salem collections from the Hawaiian Islands, Fiji, the Marquesas, Samoa, and others, are unrivaled in this country for variety and the value of the individual specimens.

Thus the handiwork collections of Salem museum, although its early growth was completely haphazard, through the fortunate circumstances which I have recited can hold its head high in any company. The same good fortune did not attend its collections of objects of natural history, most of which deteriorated until they became impossible to inflict on the observation of even the most long-suffering visitors. So also what was once considered interesting, nay even beautiful, by the standards of other days became impossible as taste changed through becoming accustomed to improved methods of preservation.

None of the arts has made greater advances during the last fifty years than the art of taxidermy. For this reason the Museum finally decided to confine its natural-history exhibitions to those setting forth the fauna of Essex County, Massachusetts, and certain special exhibits, domestic animals and ethno-zoölogical displays, for which it had unique material capable of attractive arrangement.

I do not want it to be thought that all of the natural-history material brought to Salem by the members of the East India Marine Society became completely worthless as time passed. This was by no means the case. A few objects, while they were of little interest for exhibition purposes according to present-day standards, nevertheless had a genuine scientific value. These the Salem Museum has sold to the Museum in Cambridge. There was a fine skull of the extinct California Grizzly Bear, shot in 1860. Indian Tiger skulls with definite data are not abundant in American museums, and we found in Salem the skull of what is probably the first tiger ever killed by an American – James Higginson, a great uncle of my wife's, who went to Mirzapore many years ago. Letters which tell about his hunting this very beast are still in the hands of the family. One of his hunting excursions took him to Bidar, seventy-five miles northwest of Hyderabad, whence this skull came. I have, by chance, a little scrap of paper which was found with the skull, inscribed by him "An Example of Oriental Eloquence." He went on to say:

> This beautiful petition was addressed to Warren Hastings by an Indian princess in favour of her husband who was condemned to die by that ruthless governor – it had no effect however with Hastings.
>
> "The lonely and humble slave of misery comes praying for mercy to the father of her children –
>
> "Most Mighty Sir –

"May the blessings of God ever shine upon thee; may the Sun of Glory shine round thy head; may the gates of pleasure – plenty – and happiness be ever open to thee and thine – may no sorrow distress thy days – may no grief disturb thy nights – may the pillow of peace kiss thy cheeks, and the pleasures of imagination attend thy dreams – and when length of years shall make thee entirely disengaged from all earthly joy, and the curtain of death shall gently close round thy last sleep of human existence, may the angels of thy God attend thy bed – and take care that the expiring lamp of thy life shall not receive one single blast to hasten its extinction."

To return to strictly zoölogical matters, I may add that we found that the shells of some species of tortoise have proved of considerable interest, pointing out, as I have remarked elsewhere, the giant size reached a hundred or more years ago by species which still exist today as much smaller animals.

A pair of crested domestic fowl from Java, brought back in 1826, indicate that a crested breed had already been developed in the East Indies over one hundred years ago and probably provide the stock from which such breeds as the Polish fowl and Houdan have been derived. Another big, rangy, red domesticated fowl is a really historical bird. It was brought back by Captain Richard Wheatland in 1846 from Malaya as a fighting cock. Later it was used for breeding purposes, and it is definitely known to have played a part in giving rise to the Rhode Island Red so popular and so widespread in New England today.

With this nucleus of domestic fowl already in possession, the Museum decided to do something which the Museum of Natural History at South Kensington in London has done with conspicuous success. This is to mount a collection representing all possible breeds of domesticated poultry, using, however, only the finest prize-winning examples. Thus the standards for which breeders are aiming today go on permanent record. The London collection has been in existence long enough so that changes in style affecting the various breeds have become obvious, and for this reason the collection has a real historical interest which will increase as the years roll along.

A well-known judge of poultry, Mr. Henry Pratt McKean, has already secured some splendid prize-winning birds. At Salem we plan to have exhibits, on a smaller scale, of some of the other domesticated birds, more especially the turkey and the Muscovy Duck, to show how relatively conservative these species have been in domestication, and they have been in man's hands for a very long time. We wish to point out in contradistinction the fact that the fowl, *Gallus bankiva*, has been the most plastic animal, either bird or mammal, which has come under man's influence.

There is in the Museum a codfish taken in Swampscott Harbor in 1854 by Professor F. W. Putnam when he was a young man. This fish weighed no less than eighty pounds. No such cod today could be found on the Grand Banks, to say nothing of the inshore waters of the Massachusetts coast. A giant tuna, no longer fit for exhibition, came from Beverly Harbor. It weighed some six hundred pounds and was caught in 1846 by Dr. Henry Wheatland. No tuna has been found inside Massachusetts Bay for many years, although they strike in north of Cape Ann and are abundant off the northern shores of Essex County and the coast of New Hampshire during the latter part of each summer.

The collection of Essex County birds is first class, from every point of view. The specimens are well mounted, and the number of unusual records in the form of stragglers from other parts of the country make this one of the most outstanding local collections of birds in the country, if not the very most outstanding.

The Boston Museum of Fine Arts stands first in the world in possessing the most excellent representation of objects of Japanese art to be found in or out of Japan. So also the Museum in Salem is equally noteworthy for its collections exemplifying Japanese craftsmanship. All who have read the life of Edward Sylvester Morse know of his early teaching at the University of Tokyo. Morse was a pack rat, an omnivorous, untiring, self-sacrificing, and most observing pack rat. Had it been possible, he would have collected and brought to Salem the latrines of Japan instead of merely describing them in a detailed, beautifully illustrated publication. Being of slender means, he had no opportunity to collect porcelains, but he collected everything in the shape of pottery which Japan produced. His great definitive collection is in Boston, his second collection is in Salem. There, also, he gathered, with the aid of his devoted and utterly charming friend, Dr. Charles G. Weld, everything else which the Japanese use on their daily occasions — carpenter's tools, writing materials, the myriad knots tied for ordinary and special occasions, vehicles, firemen's uniforms, actors' gear, — he assembled a great series of collections which is absolutely unique. There is nothing like it to be found in Japan or out.

This fact has been stated over and over again, as the Museum in Salem was visited by every noteworthy Japanese who in the days before the war came to this country. This institution is therefore a monument to Morse as well as to the East India Marine Society. The beautiful hall in which these objects are housed is appropriately called Weld Hall. The old original building, once the offices of the Asiatic Bank and the Oriental Insurance Company, has recently been renovated, and the original charm of this fine structure has been restored.

The lovely hall in the second story of the building, a room of superb proportions and charming design, had been filled with a mass of unsightly cases and cluttered with dreary galleries which housed not only the Essex County Natural History collections but the hodgepodge accumulated through the years to form the so-called natural-history section of the Museum. Last year the trustees decided to clear out all this welter of casing, and the hall today, with the great figureheads, large ship models, and other monumental objects set where you may see them to the best advantage, is uniquely impressive and beautiful. Of course, Salem had its glory holes. An attic over the great hall, reached by means of a trapdoor and a ladder, contained thousands upon thousands of mouldering herbarium specimens untouched for many decades. Fortunately these were recovered before they were completely worthless and are now in herbaria where they will be properly cared for. It was as if the behest of Caleb Cooke made in February 1857 had been applied to many of the objects of the Museum. Mr. Cooke wrote on the paper wrapping a parcel of shells, "Please do not disturb these shells." This behest was scrupulously obeyed for eighty-five years and six months, when in 1943 the iconoclastic writer of these lines began to work on the glory holes of Salem.

UPPER: THE MOST DISTINGUISHED EXHIBITION HALL IN AMERICA:
PEABODY MUSEUM, SALEM, 1944
LOWER: THE SAME HALL AS IT USED TO APPEAR

THE MUSEUM OF COMPARATIVE ZOOLOGY AS ORIGINALLY PLANNED AND AS IT LOOKS TODAY

CHAPTER V

Rare and Historic Birds

ON the evening of March 17, 1930, the Nuttall Ornithological Club celebrated the fiftieth anniversary of Outram Bangs's election to membership, and at that time his paper on "Types of Birds now in the Museum of Comparative Zoölogy" was handed around just from the press. This listed holotypes or cotypes of 1241 species or subspecies of birds representing the work of no less than ninety-three authors, either publishing alone or jointly. Surely this is an astounding record, considering that most of the material represented had been secured within the previous twenty-five years!

Since that occasion Bangs's successor, James Lee Peters, has continued an open manuscript list of types written partly by Outram before his death and continued by Peters. This lists 230 types in addition to those in Bangs's first list, so that at the present time there are 1471 all told. In the recent list, naturally, the names of Peters, Griscom, and Greenway appear most frequently. The name of Professor Oscar Neumann is there frequently also, for from him we have purchased a very considerable number of the original specimens of birds which he has described both from Africa and southeastern Asia.

I am often asked how many specimens of birds there are in the Museum in Cambridge. This question I cannot answer, nor can anyone else, without spending months of time and trouble which is not worth the doing.

We have received five great collections of North American birds within, relatively, a few years of each other, so that inevitably there has been some accumulation of duplicate specimens on our hands. Each of these collections, nevertheless, fortunately supplemented the rest. Mr. Brewster's was especially strong in birds from Lower California and parts of Mexico; Mr. A. C. Bent's collection, in early plumages; Mr. F. H. Kennard's was rich in waterfowl, especially geese. The great collection which John Eliot Thayer left us was well rounded and replete with rare and extinct forms, while Mr. Charles F. Batchelder's collections contained not only a number of types and representations from a number of relatively little-known regions in the western part of North America but also from Newfoundland.

The result is that the combination of this material has given the Museum a collection of North American birds probably only excelled by that in the possession of the United States Government. As I said before, we have found ourselves with duplicates on hand, and since it is not our policy to collect birds to make a showing of numbers, individuals have been disposed of in many directions. A guess as to the number of birds now in the house would inevitably amount to nothing.

Visitors to the Museum are always interested in our two beautifully mounted Great Auks. One of these was a gift of my father to the Museum and the other was a gift of John Thayer. He also had in his collection enough Great Auk bones to make up many composite skeletons, as well as a round dozen of Great Auk's eggs, of which the Museum already had one very fine specimen likewise received from my father.

Our Labrador Ducks are three in number, fewer than those in the possession of the museums either in New York or in Philadelphia, but they are quite unique in that two are beautiful study skins as fresh as if they had been made yesterday, while the third is a superbly mounted adult male which Mr. Thayer secured from another private collection many years ago. Our two skins came from Mr. Brewster, and one of them has a very interesting history. Traveling through from the West to Boston, he was forced to change trains in Albany. He had some time to spare, and, having a toothache, he inquired at the office of the hotel where he found himself at the moment for the name of a dentist. He sought out the dentist at his office and beheld, standing on the mantelpiece, mounted under a glass cover, a bird which he knew at once to be an immature male Labrador Duck. Mr. Brewster explained, for he was most honorable in such matters, the rarity of the bird which he had seen. The dentist was taken completely by surprise. A good price was set, which Mr. Brewster doubled. He then had his tooth extracted and all transactions were satisfactorily concluded. Mr. Brewster already had a female bird in his collection, so that the Museum now possesses an adult pair and this immature male, an entirely adequate representation of this rare bird.

Mr. Thayer's collection was also particularly rich in representation of the birds of Guadeloupe Island, off the Pacific coast of Lower California. Here, because of the introduction of dogs and cats and goats, all of which ran wild and multiplied exceedingly, a number of species of birds became extinct. Specimens of each one of these are now in the possession of the Museum and some are among the rarest of the extinct birds of the world. By this I mean that they are represented only by very few individuals.

I don't intend to turn this little essay into a list of the extinct and historical birds in the Museum. Nevertheless I'm always rather thrilled when I pass through the synoptic gallery of birds and see the bald eagle which was actually mounted by Alexander Wilson, and exhibited alongside of it the drawing he prepared from this very bird for his great *American Ornithology*.

One of our specimens of Parkman's wren is the very bird mounted by Audubon and given by him to Francis Parkman, for whom he had named the species.

Parkman had this bird in a little glass box on his desk until the day of his death. Then Mr. Thayer got it and gave it to the Museum. If you're a sentimental old codger, as I admit perfectly frankly that I am, a specimen such as this one carries an associational value which is out of all proportion to its worth purely as a zoölogical specimen.

I spoke some time ago of the vicissitudes through which Wilson's birds passed before they came into our hands. Most of the early American museums of natural history were proprietary, really commercial institutions. Most famous among these was Peale's museum at Philadelphia.

Charles Willson Peale was born in Chestertown, Maryland, in 1741. He started his museum in 1784 in a frame building annexed to his dwelling at the southwest corner of Lombard and Third streets in Philadelphia. In 1794 his collection was moved to the hall of the American Philosophical Society, and in 1802 the State of Pennsylvania granted the use of a part of Independence Hall for the exhibition of his material. The active management of the museum fell upon the shoulders of Peale's sons in 1808, and in 1820 the museum was incorporated as a stock company. In 1828 a move was made which placed it in the Arcade on Chestnut Street above Sixth Street. In 1838 it was again moved to a building at Ninth and Sansom streets. In 1846 the Philadelphia Museum

Company came to grief, and the collections were all sold at auction.

The natural-history material, however, was kept together and continued to be displayed in Masonic Hall in Philadelphia until 1850, when it was purchased by Moses Kimball and P. T. Barnum. Barnum's half of the material went to his American Museum in New York and was completely destroyed by fire on the 13th of July, 1865. Kimball had in 1839 become the proprietor of the New England Museum, and in 1841 had removed the material of that institution into the Boston Museum, which he also owned. This was located first on the corner of Tremont and Bromfield streets. Afterward, in 1846, it was still in Tremont Street, but moved down towards Court Street.

Here in 1850 Mr. Kimball brought his half of the Peale Museum material. This information has been derived from Doctor Walter Faxon's informative article concerning the "Relics of Peale's Museum" which was published in 1915 in the Bulletin of the Museum of Comparative Zoölogy. Faxon adds, "The Peale Museum was the repository of a large number of the types of species described by C. L. Bonaparte, Richard Harlan, George Ord, Thomas Say, and Alexander Wilson." Thus a careful examination of any material which had been in Peale's hands might reveal specimens which had great historical interest. This interest had an added potential importance because many of the most prominent naturalists of Europe — Geoffroy Saint Hilaire, Cuvier, Lamarck, John Latham, and Maximilian, Prince of Wied – were also correspondents of Peale's. There is a tradition, too, that parts of Sir Ashton Lever's famous collections, dispersed by public sale in London in 1806, had found their way into Peale's museum.

The booty of the expeditions of Lewis and Clark and Major Long as well as other early historical explorations made the Peale Museum a veritable treasure house, as it became the depository of a large number of the types of animals described by the Philadelphia naturalists of the early days. Among the most treasured possessions of the Peabody Museum at Harvard University are a number of objects that were secured by trade from the Indians of the Northwest Territory and brought back by Lewis and Clark. There are some early porcupine-quill-worked moccasins and tobacco pouches made of otter hides which are superb examples of early Indian workmanship. The famous wooden bowl in the shape of a beaver which was also once in the possession of the Peale Museum and which came into its hands from Judge George Turner of the Northwest Territory is a treasure worthy to stand as it does beside King Philip's samp bowl and the Sudbury Bow, which was preserved by William Goodenough "who killed the Indian in 1660." This, incidentally, was the only Massachusetts Indian bow available, and was used as a model when the State shield was designed.

Part of the Peale Museum was given to the Boston Society of Natural History in 1893, the residue in 1899 after the Boston Museum building was damaged by fire in May of that year. The history of this priceless material after it fell into the hands of the Boston Society of Natural History is a shocking tale of shameful incompetence. Some of the specimens were forthwith destroyed and the others were sold in 1900 to Mr. C. J. Maynard, who sent the collection to his residence in Newton, Massachusetts, and stored it for a while in his barn. It was subsequently repurchased by the Society. The birds were wrenched from their stands and packed into tin cans, when in reality they never should have been touched until a competent person had studied each specimen carefully. In many cases, it is known that the original labels were pasted fast to the bottom of the stands on which the birds were mounted, hence these were lost when this step was taken. The next move was one which it would be more pleasant to pass over, were it not that a historical fact is involved which explains a great deal. A person of whose intelligence the less is said the better was employed – mind you I say employed – by the Society to examine this collection and report upon it. He removed all the remaining original Peale labels, many of which were pinned or tucked under the wings of the individual birds. These labels he proceeded to place in a paper envelope and lose.

The only surviving labels which go back to the

UPPER AND MIDDLE: EXAMPLES TO SHOW HOW PROFESSOR PECK DRIED HIS FISHES

LOWER: COTYPE OF DREPANIS PACIFICA

OLD EXHIBITION CASES IN THE M.C.Z.

Peale Museum are two wooden ones belonging to a pair of Golden Pheasants presented to Charles Willson Peale by General George Washington. Doctor Faxon remarks that Mr. Maynard told him that he remembered two groups of mounted birds, arranged in two glass cases, presented by Washington to the Peale Museum. These were transferred with the rest of the Boston Museum to the rooms of the Natural History Society, but Maynard said that they had been disposed of before he purchased the collection. Maynard was in error. One of the groups evidently was disposed of, and concerning its disposition we have no inkling. The Pheasants, however, beautifully remounted by George Nelson, are in this Museum now – one of the most treasured and interesting of our exhibits.

Now in spite of the fact that all the historical information concerning the Peale Collection was known, not a single specimen could ever have been satisfactorily identified had it not been for one single well-established historical fact.

As Faxon tells us, Charles R. Leslie, in his *Autobiographical Recollections*, says of Alexander Wilson, "He worked from birds which he had shot and stuffed, and I well remember the extreme accuracy of his drawings, how carefully he counted the scales on the tiny legs and feet of his subjects." Wilson had no training as an artist, but worked obviously with the most scrupulous care. This is made clear not only from such original drawings of Wilson's as are now in our possession but from the illustrations in the *Ornithology* as well. Hanging framed on the wall of my office are the original water-colors for five of his plates, and the museum has also many others representing groups painstakingly assembled and copied from badly mounted birds.

I spent many pleasant hours with Doctor Faxon while he worked over the Peale birds, one by one. There were many, many cases when it was to be presumed that the bird, possibly a type or a figured specimen, had been in Wilson's hands, but in the case of common birds, mounted in a commonplace way, proof would be impossible to establish. Every once in a while, however, a specimen would turn up where the evidence was indisputably clear. For example, in the case of the Black Vulture, the mounted bird has several white feathers by chance in one of the wings, and thus it was figured. In a number of other cases the birds were mounted in peculiar and unlikely postures to make them fit into the space available on one of the usually crowded plates. Here the identity could be definitely established. The upshot was that, with three Wilson birds, one in the possession of Vassar College and two in the Academy of Natural Sciences in Philadelphia which were already known, Faxon established clearly that no less than 52 other individuals in the possession of our Museum were unquestionably figured specimens and in many cases types as well. These now have been carefully segregated from the rest of the sorry welter of zoölogical trash left after inexcusable bungling.

Some years ago I interested Mr. James C. Greenway, Jr., in the idea that it might be possible to establish the identity of the Peale Museum specimens which had been purchased at the dispersal of the Leverian Museum in England. The specimens were all reëxamined with the hope that some other clues might reveal their identity, but this was not the case. Wilson had his curious habit of drawing directly from a mounted specimen which made some identities certain; Audubon, you will recall, drew from the fresh bird. Wilson's custom made it possible to salvage a tiny modicum from the great number of specimens which, being of no further use, have been discarded.

The history of the Washington Golden Pheasants I had the good fortune to discover by reading the General's diaries. He wrote, on Monday, November 27, 1786, that he had received a cock and hen Golden Pheasant as a gift from his friend the Marquis de Lafayette. Only a year or two ago I had the pleasure of learning that as a result of this information the Board of Lady Governors at Mount Vernon were planning to exhibit a pen of living Golden Pheasants at the location where it is quite probable that Washington kept these birds for such time as they lived. That they were kept in a cage I believe to be certain from what we know of the habits of the Golden Pheasants. These birds are not of a nature to permit their being set free around the spacious lawns of Mount Vernon and expected to remain in the neighborhood. The fact that our male and female were recovered and saved when they died, I believe, proves that they were kept in a cage. The remarkable thing is that, in spite of time that must have elapsed before they reached Peale's hands after their death, they are in fine plumage and excellent condition. This suggests that they lived in captivity but a short time and died during the cold weather of 1786–87.

CHAPTER VI

The Swing of the Pendulum

THE science of zoogeography may be said to have been born in 1876 when Alfred Russel Wallace wrote *The Geographical Distribution of Animals, with a Study of the Relations of Living and Extinct Faunas as Elucidating the Past Changes of the Earth's Surface*. Four years later Wallace published an amplification of certain aspects of the subject matter of this earlier volume under the title *Island Life; or The Phenomena and Causes of Insular Faunas and Floras, Including a Revision and Attempted Solution of the Problem of Geological Climates*. These are the two great classics upon which zoography rests as a foundation. I confess that I prefer the word "zoography," rather than the word in more common use which is "zoogeography." It slips off the tongue a little more easily and serves its purpose equally well.

Previous to Wallace's writing, an enormous literature had sprung up outlining the general facts of animal distribution, but no one had paid very much heed to *how* animals and plants were dispersed until Wallace appeared on the scene.

It is a truism that if a pendulum swings one way, sooner or later it swings back again, and nothing is more certain than that this has happened with the branch of science under discussion. It is difficult to say who was responsible for the first plethora of postulated land bridges which in bewildering numbers were called into being to explain the distribution of animal life as we see it today. The consequence was, however, that it became easy and convenient to account for the distribution of animals by means of temporary land bridges, and there is no question that some absurd deductions resulted. I remember once to have heard a statement, if I did not actually read it in print, that the presence of *Eumeces longirostris* on Bermuda was only explainable by the fact that Bermuda was once connected with North America. A connection of Florida with Cuba was also considered to be not improbable, and Lemuria, that hypothetical land across the Indian Ocean, and Gondwanaland, connecting South America with West Africa, were generally invoked to account, for instance, for the distribution of the fresh-water Characine fishes. Of course the first two of the examples which I have mentioned are admittedly absurd. The other two are based on a more reasonable conjecture. However, it is a fact that if such land masses ever did exist it was undoubtedly so long ago that they played no part in explaining the distribution of vertebrate life.

Let us take for example the history of opinion with regard to the Greater Antilles. Wallace was of the opinion that the Greater Antilles had been connected one with another and with the mainland. Robert H. Hill, writing in 1895, inclined to the view that there never had been any great change in the distribution of the land masses from the condition in which they are found to be at the present time. In 1916,* Doctor Matthew and I published an attempt in a very friendly way to set forth impartially the opposite points of view. Matthew disliked to call into being any land bridges to explain existing conditions, while I maintained the opposite attitude. To be sure Matthew did admit later the possibility of a connection between Jamaica and Haiti during the Tertiary, but, in 1906, he held that the Greater Antilles were oceanic islands, and he rather strengthened his conviction in 1915.

I still maintain the essential probability of the views which I then expressed, namely that there was a long-standing connection between Central America and Jamaica as well as connections between Jamaica, Haiti, Puerto Rico, and the Virgin Islands, and another between Cuba, Haiti, Puerto Rico, and the Virgin Islands. These "land bridges," if you wish to call them such, are – generally speaking – found to be quite acceptable to many geologists, as well. The paleontological evidence which has come forward recently substantiates the probability of the continuity of the land.

* W. D. Matthew, "Climate and Evolution," with supplementary note by Barbour, *Annals New York Academy of Science*, 27:1–15.

LIBRARY
LOS ANGELES COUNTY MUSEUM
EXPOSITION PARK

There is less unanimity of agreement concerning the Cuba-Yucatan connection which I once considered to be relatively certain and of long duration. Schuchert, in his *Historical Geology of the Antillean-Caribbean Region* (1935), believes that the Cuba-Yucatan connection was a very early one and that there is nothing to indicate a connection during the Tertiary. I am perfectly free to confess that I am unable myself competently to evaluate geological evidence. I have perhaps in times past been too free in assuming changes in land forms. However, in many conversations with the late Professor William M. Davis, my friend Doctor Reginald A. Daly, and the late Doctor Charles Schuchert as well, I find that geologists do not shudder at the assumption that great changes may have taken place in the past distribution of land, and they are much more liberal in their views than many zoölogists. There is a growing tendency, on the part of many zoographers, to explain the distribution of all life everywhere on the assumption that migration took place, without calling into being any non-existing land masses, while geologists often acquiesce in the assumption of the changes.

My own conviction in regard to all these matters is like many other personal convictions – variable. I know perfectly well that I have unquestionably made mistakes in assumptions which I have used in zoogeographic explanation. However, I cannot bring myself to agree with my talented young colleague, Philip

Darlington, for instance, or the learned Dr. Ernst Mayr, who would explain the transportal of fresh-water fish, amphibians, in fact pretty much anything or everything, to and including Peripatus, to fortuitous flotsam-and-jetsam dispersal, relying especially on hurricanes. These I suspect actually to be harsh, rough, and fearfully desiccating for many delicate organisms.

I don't want to imply that the stage has been reached as yet when the geologists will have to defend their conclusions regarding land connection against the rising tide of isolationist zoölogists, but the zoölogical pendulum is certainly at present swinging sharply toward the isolationist point of view.

To be sure there seems to be a general acquiescence among all groups in favor of the glacial-control theory brought forward originally by Professor R. A. Daly in the *American Journal of Science.** This he later amplified in the Proceedings of the American Academy of Arts and Sciences.† I imagine that the fundamental outline of this idea may be so generally and widely known that no amplification is necessary, but perhaps for the benefit of laymen who may not be cognizant of the theory but still be interested, I may say that this presupposes a lowering of from 200 to perhaps as much as 300 feet in the water level of the oceans. The cause for this is the withdrawal of water

* 4 ser. 30 (1910).

† 51:158–251 (November 1915).

from oceanic distribution to form the polar ice cap during each one of the periods of maximum glaciation. This would account for a drying up of the shallow oceanic areas and provide a simple and extremely plausible explanation for the connection of Siberia with Alaska, of England with the continent of Europe, of Sumatra and Borneo with the mainland, of New Guinea with Australia. In the West Indies, the region in which I am particularly interested, this would make an enormous great spidery land mass, large but of very irregular form, out of the Antillean Archipelago and the Bahamas.

What caused these glacial periods is very much a moot question, on which there has never been any unanimity of opinion. Nevertheless, to my own way of thinking, Professor Harlow Shapley has set forth by far the most plausible explanation of any which I have ever heard. He believes that the earth has passed in the course of its movements in space through the tail of an enormous comet. This was a phenomenon which took place on a scale so vast that the temperature of the atmosphere was profoundly altered for a considerable period of time, the loss of heat from the earth by radiation being greatly reduced. This theory certainly offers a possible explanation which is well worth our serious consideration. Unless a similar phenomenon happened perhaps on repeated occasions, many millions of years ago, some other explanation must be sought for the climatic changes which took place in carboniferous times – when Greenland grew palms, as the fossil evidence proves it did, and there was vegetation on the Antarctic continent.

The whole matter which I have been discussing was brought forward in my mind when I received a most absorbing paper by Doctor Ernst Mayr entitled "The Birds of Timor and Sumba" (*Bulletin of the American Museum of Natural History*, 83:129–194, 1944). Mayr's attitude of mind is summed up when he says, "A good part of historical zoogeography will have to be rewritten. Land bridges have been erected in the past with such utter disregard for geological ecological, and phylogenetic data that almost every land bridge is now under suspicion." I would hasten to add the words, "except those which the geologists *insist* upon as being certainly proved," and, fortunately for the peace of mind of the zoographical tyro and skeptic like myself, there are a very great many of these.

Mayr continues: "In a way it has been a great disappointment to the zoogeographer to realize how extensive the dispersal faculty of animals and particularly of flightless animals is. Since it has become obvious that the present distribution of animals is only a poor clue in the analysis of every land connection it might even be asked whether regional zoogeography can contribute anything at all to facilitate paleogeographic reconstruction." Doctor Mayr goes on to speak of the well-balanced continental faunas of

Formosa, Borneo, and New Guinea. He maintains that these prove the islands to have been in recent connection with the continents, although just where and how New Guinea was connected with Asia, or when, is not defined. In any case I am certainly entirely in agreement with the general views.

I feel less absolutely sure, however, that the same definite statement can be made that Celebes has never had any continental connection. The separation from Borneo, although narrow, is beyond a shadow of a doubt very ancient, geologically speaking. The land connection with Java, once considered to be quite probable, perhaps has never been in existence. Let us admit this; but the connection with the Philippine Islands is to say the very least one which cannot be cast aside as being absolutely impossible.

In August 1883, a terrible eruption of Krakatoa took place, and I quote the following, written by my friend Dr. K. W. Dammerman of the Buitenzorg Museum (*Treubia*, vol. III, 1923, p. 61).

> The terrible eruption of Krakatau, August 1883, has been such an unexpected opportunity for biologists as perhaps will not occur for years. Although the consequences of the disaster were dreadful, no less than 30,000 people having perished, this experiment of Nature is of most interest for zoogeographical problems, especially how a barren island, wholly destitute of animal life, is reoccupied again.
>
> The question about the total devastation of the

fauna of the islands of the Krakatau-group in 1883 cannot be settled absolutely, but there is every evidence that no animal could have survived the eruption. From the 20th May till the 26th August the explosions followed each other with short interruptions, covering the islands with stones and ashes. By the last and most violent explosion the volcanoes Danau of 450 M. in height and Perbuwatan disappeared altogether, and the Peak of Rakata or Krakatau of 800 M. was split in its very midst and one half blown away. The three islands which remained after the eruption, Krakatau (Krakatoa), Verlaten Island (Forsaken I.) and Lang Island, were overshed by hot ashes, a layer of 30–60 M. thickness! No animal could have remained in his hiding place during the explosions and, buried by the ashes, it could not have escaped from destruction, the bottom layer of ashes remaining hot for days. The possibility that a single animal, concealed in a recess of the rocks, survived the disaster may be maintained, but such an animal would have perished after a short time as no food was available, the whole vegetation also being destroyed or burnt. All biologists who have visited the islands after the eruption are of the same opinion. But granted to sceptics that a single animal did survive and could maintain itself after the eruption, this is of little or no importance, for certainly 99% of the animals living now on the islands are new invaders.

After years had passed, for Dammerman's account appeared in 1922, C. A. Backer published a book in December 1929 entitled *The Problem of Krakatoa as Seen by a Botanist.* In this book Backer, who is obviously a very competent but a very contentious person, finds fault with the conclusions of almost everyone who has visited Krakatoa since the eruption but his own. Nevertheless, the argument as to whether or not Krakatoa was completely sterilized botanically is one thing. Whether or not it was sterilized zoölogically is quite another, and the point of the whole matter is that the types of animals which have reinvaded the island are just those which would have been supposed to be the first to appear. Geckos and skinks are among those lizards most frequently carried about by man, or in any event, most widespread on islands where there is absolutely no question of any possibility of the islands ever having had any connection with the mainland.

I suspect that in the case of Krakatoa large lizards such as Varanus, and pythons, swam there, for they are known to be fearless and at times swim out to sea, and I am quite sure that the shores of the island have been the scene of many fishermen's camps, and lizards of the two families which I have mentioned are no doubt very often stowaways.

Oceanic islands, and here the extreme isolationists among my friends I know will agree, can sometimes be spotted at once so that all hands are in complete agreement. Look at the faunas of the Gilberts, the Marshalls, the Paumotos, the Hawaiians, and a myriad other islands in the Pacific and there will be no chance for an argument. Come to Fiji, however, and the situation is quite different. So it is also in the case of the Sunda Islands, where land is known to be in unstable equilibrium and where the changes elevating one area and depressing another have certainly been very great. Here islands with an impoverished fauna may have had a large part of their population wiped out by submersion, and the remnant is quite likely not to give a clear picture of what the original fauna of the island really was. I believe that this states one of the pitfalls in which the isolationist may find himself.

As I got to thinking the matter over, it occurred to me that there was one who knows this Malaysian area from wide travel and observation, and who has given this whole question great thought. I spoke to my friend and colleague, Professor E. D. Merrill, and he said that he was writing something along the general lines of the distribution of the plants and animals of the Malayan and Papuasian islands and that I might quote what he had written.

Admitted that zoography and phytogeography are by no means to be considered as being one and the same science, nevertheless, as concerning the matter of land connectings, the real matter under discussion, they have this in common: both animals and plants will distribute themselves by land if it is there, as well as by flotsam and jetsam and by hurricane if it is not. Here is what Dr. Merrill wrote me:

> As I see the picture of the geographic distribution of plants and animals in the Malaysian region there are apparently two great centers of origin and distribution within the Archipelago as a whole, these being the Sunda Islands in the west and New Guinea in the southeast. The Archipelago lies in a very strategic position in relation to the problems of the relationships of the Asiatic and Australian faunas and floras. To a marked degree the Asiatic (continental) types of both plants and animals tend to become less dominant as one proceeds to the south and east, and likewise the Australian elements become more and more attenuated as one proceeds north and more particularly to the west.
>
> I visualize two more or less stable areas; that is, reasonably stable since the close of the Tertiary, although in both there seem to have been some elevations and depressions. These are the areas designated as Sundaland, consisting of all the Sunda Islands, including the Malay Peninsula and the Palawan-Calamianes group in the Philippines, and Papualand, consisting of New Guinea and adjacent Islands, both delimited by the present continental shelves, the 200-fathom line. Between these more stable areas is an intermediate region consisting of the Philippine Islands (except the Palawan-Calami-

anes group), Celebes, Gilolo, the Moluccas, and the Lesser Sunda Islands as far west as Lombok. In this unstable area great elevations and depressions seem clearly to have been the rule since the close of the Tertiary, with intermittent land connections here and there permitting certain intermigrations of both plants and animals. But such land connections as there may have been seem in general to have been north and south within this area rather than east and west. There is little or no evidence of direct land connections between Asia and Australia since the close of the Tertiary.

Contrast this intermediate unstable area with Sundaland to the west and Papualand to the southeast. This insular region lying between Wallace's and Weber's lines (both modified) has been designated as Wallacea. It has a very irregular submarine topography with more or less evident rows of high islands alternating with various great deeps, some as much as 4,000 to 5,000 meters deep. Really these are "some holes" when one contrasts this unstable area with the shelf areas of Sundaland and Papualand wherein deeps are absolutely lacking!

The continental shelf areas have a remarkably even submarine topography, the average depth of the water covering them being but about 100 feet. The Asiatic shelf is about 1,500 miles wide. No great deeps occur anywhere within the limits of these great shelf areas. A change in sea level today of only about 150 feet would unite all of the Sunda Islands and the Palawan-Calamian group with continental Asia, and one of 75 feet would unite New Guinea with Australia. Attention is called to the fact that the fresh-water fishes of the *northern* Borneo rivers are practically identical with those of the *eastern* Sumatra rivers, and that tin dredging is now carried on far out to sea in the old drowned channel of the once-great river that flowed to the northeast, of which the northern Borneo and eastern Sumatra rivers were once tributaries.

In the Philippines, and I believe also in the islands immediately to the south, are various conspicuous benches, marking the positions of ancient seashores. In some places one may note a series of these, some of them at least as much as 4,000 feet above the present sea level. The last great episode in the history of Celebes and of Gilolo [Halmahera] seems to have been a great depression, leaving only the mountain systems that form the skeletons of these islands above the sea, and drowning the great valleys. Sink Mindanao 100 to 150 feet and the result would be a replica of these two islands, for the great valleys of the Cottabato and Agusan rivers would be flooded for much of their length. Even today on the Agusan, in times of heavy rainfall on its lower tributaries, the river current changes and flows strongly inland, flooding the great marsh area on its upper reaches.

Being a botanist and not a zoölogist, I do not believe in sharp lines of demarcation as many zoölogists seem to do. Rather I consider Wallace's line as modified to be the eastern boundary of an ancient continent, and Weber's line, also modified, the northern and western boundary of another ancient continent. Please note that we have modified the position of Wallace's line, extending it northward through the Sibutu Passage, the Sulu Sea, the Mindoro Strait, thence along the western coast of Luzon and into the great deep east of Formosa between Formosa and the small island of Botel Tobago. Note that Botel Tobago contains strong Philippine elements of plant life, and that these scarcely extend to Formosa proper. I modify the position of Weber's line, placing it close to the New Guinea continental shelf and extending it into the Pacific Ocean between the western tip of New Guinea and Gilolo.

I consider the flora and fauna of this great intermediate area to be made up of relic species of its original bios, and their descendants, plus infiltrations from both the west and the north as well as from the south and east. A considerable number of Australian faunal types reach the Philippines, and a very large number of plant types. However, very few of these Papuan-Australian plant types have reached the Sunda Islands, and also I judge few animals. In botany, as we get deeper and deeper into the New Guinea flora, the number of types previously known only from the Philippines increases radically. Intermigrations have certainly occurred, and it is difficult to explain these without postulating some land connections. East and west migrations within this unstable area have not been strong, but north and south ones are striking. In general it seems to me that migrations from the Sunda Islands have been *northeastward* through Palawan and Sulu into the Philippines, and thence *southwestward* to New Guinea; and in reverse the Papuan-Australian elements extended *northwestward* strongly into the Philippines (even to northern Luzon and the small islands between Luzon and Formosa, but none reached Formosa), but for the most part failed to negotiate the northeast-southwest leg of this triangle, from the Philippines to the Sunda Islands. I, therefore, postulate a round-about migration route with considerable confidence that it is reasonably correct; this is definitely true for plant life, but it may or may not apply to animal life.

Please note the position and character of the Sangir Islands, an extension south in the direction of the Minahassa Peninsula of Celebes, being a tapering off of the central Mindanao mountainous region, and also the scattering islands between the southeast corner of Mindanao and Gilolo, the drowned remnants of the Mindanao east-coast mountain chain.

Taking up elevations and depressions, in Luzon fossil plant remains at 5,000 to 6,000 feet can be absolutely matched with low-altitude living species none of which now occur more than 2,000 feet above sea level. Marl beds near Baguio at 5,000 feet contain shells that are replicas of those living on the adjacent coast of eastern Luzon.

In studying present-day distribution of plants and animals, neither hydrography nor geological history can safely be ignored. Many existing islands have not always been islands, and from what we now know of Malaysian geologic history we may, with reasonable safety, postulate two more or less stable continental areas within what is now an insular region and which has certainly been insular throughout the Recent. But between these previously existing continental areas there is an intermediate one as outlined above which has had an entirely different geologic history in Pliocene-Pleistocene times. One scarcely has to go beyond a study of submarine topography to visualize the difference.*

This seems to me a pretty good exposition of the situation. The fact that von Mollengraaf described

* See the introduction to Merrill, *Enumeration of Philippine Flowering Plants*, 4: 77–154 (1926), and Merrill, "Distribution of the Dipterocarpaceae; Origin and Relationships of the Philippine Flora and Causes of the Differences between the Floras of Eastern and Western Malaysia," *Philippine Journal of Science*, 23:1–33, 2 maps (1923).

areas composed of radiolarian ooze, beyond doubt of deep-sea origin, now at considerable elevation on the islands of Timor and Borneo would prove if nothing else did that this is an area of great isostatic instability. Of this the volcanos are another indication.

So the matter rests, and in the face of the apparent facts I cannot believe that so many of these islands have always been isolated any more here than in the Caribbean, where the conditions were not wholly dissimilar.

CHAPTER VII

Thinking Out Loud

I ONCE, perhaps happily erroneously, had the firm conviction that if I were to be completely disinterested in my desire to serve the Agassiz Museum I should permanently give up all scientific work and literary work and devote myself for such years as may perchance remain allotted to me to nothing but the preparing of descriptive labels. It has been said, time and again, that a good museum exhibit is a collection of arresting and well-composed labels illustrated by a few carefully chosen specimens. This, however, no longer seems to be as true as once it was. The other evening, sitting in the shade in front of my little house in David Fairchild's garden at Coconut Grove in Florida, I remarked to him that in the light of present experience it made me rather sick at heart to consider the amount of time I had wasted preparing labels for the museum at Cambridge, and we began to philosophize on the whole matter.

Let us take, as our first example to consider, the Thayer Collection of North American Birds. This is exhibited principally so that amateurs of ornithology who spot birds unknown to them during their field observations may come and look at our collection and thus identify the birds which they have seen and which have puzzled them. Here is a clear case where the name of the bird, English and Latin, and its range, pretty much answers the needs of the vast majority of visitors. These data are all the label needs to bear.

For a second example, consider our collection representing bird architecture. It has always been popular with the public ever since Ludlow Griscom installed it so skillfully at my request. He prepared one large general label which tells the story which the various nests shown were chosen to illustrate. They point out the general categories into which the different types of nests fall. At one end of the series you may see slight depressions kicked out of the surface of a cinder dump in which eggs are laid by the killdeer plover, while at the other extreme are shown structures as elaborate as those made by the Central American palm swift (Panyptila), a bird which constructs a long stocking-like affair woven of kapok and other fine vegetable wool. Up this tube the swift creeps to reach the tiny shelf where its eggs are laid.

It is of interest that this bird, long considered a very rare one and difficult to find in the region of Panama, is now often to be observed since it has taken to attaching its extraordinary nests to the masonry walls of the buildings controlling the locks along the Canal and to other stone structures instead of to the trunks of widely scattered forest trees in the jungle.

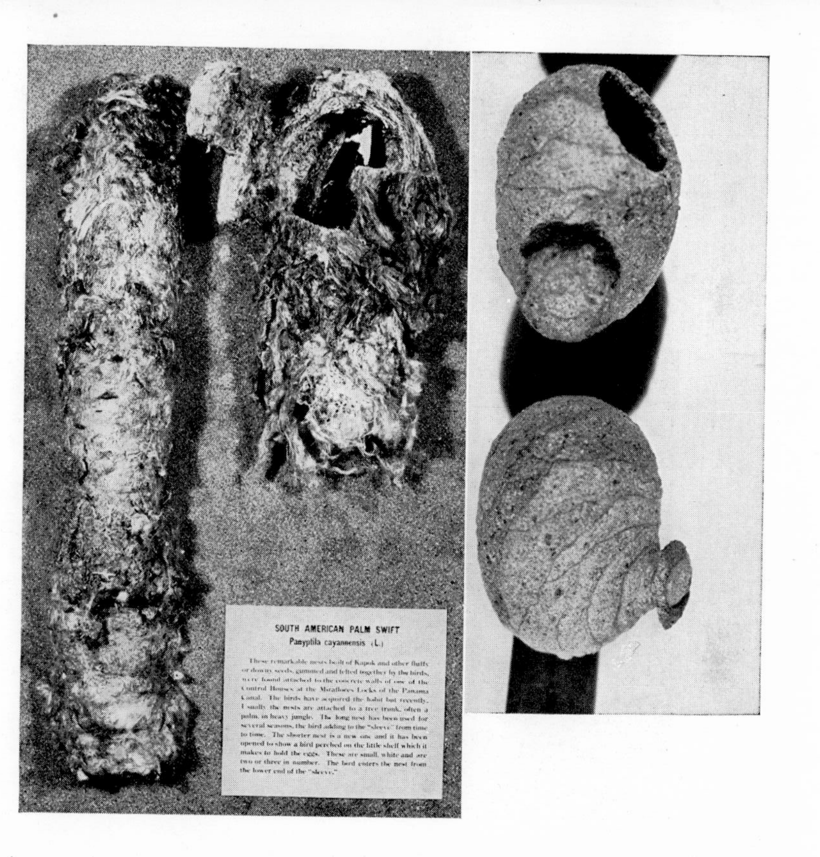

LEFT: NEST OF A
PALM SWIFT,
FROM PANAMA

RIGHT: TWO CELLS
MADE BY EUMENES,
THE POTTER WASP.
ENLARGED
ABOUT TWICE

MR. ALEXANDER AGASSIZ IN HIS STUDY AT NEWPORT

I even saw one nest being built to hang from the glass shield covering a powerful electric light over one of the entrances to the Gorgas Hospital.

Hummingbird nests always excite interest, whether made as our native species builds, a tiny cup saddled on a limb and camouflaged with bits of lichen stuck to its surface, or made as by some of the tropical genera, such as Phaethornis, where a cluster of tall grasses is in some perfectly inexplicable manner bound together into a bundle with the tiny nest of vegetable down inserted intriguingly in the bundle of grasses. This feat might be relatively simple for a larger and more powerful bird, but how it is accomplished by a tiny hummer is a mystery indeed. In this collection the individual units are inherently beautiful and why they are exhibited is quite obvious. They are examined by a considerable public who pay no attention to the labels whatsoever.

Some years ago my colleagues Frank Carpenter, Joseph Bequaert, and I set up an exhibition illustrating insect architecture. Here many of the objects shown are also quite lovely. The individual units vary from a tiny cell protecting a single egg and its emerging larva to some enormous and very complicated structures. They are varied in the extreme. None, I am sure, excels in delicacy a minute yet absolutely perfect globe of clay, shaped as if it were formed on a fairy potter's wheel. It is attached to a needlelike spine of a Panamanian black palm. This was the work of a member of the genus Eumenes, certainly a master potter among the wasps of the world.

All of these objects are unfamiliar to the layman, and this is another of the collections for which there are descriptive and explanatory labels, but these labels are no more often read than any others.

Take next the case of vertebrate fossils. Here, contrary to our usual custom, type specimens – meaning, of course, those on which descriptions of new forms are based – are put on exhibition. In the case of birds, or mammals, or reptiles, or fishes, or insects, these would all go to the study collections and be kept away from the light. Fossils, however fragile, are naturally uninjured by light, and are really much safer on exhibition than in the trays of the study collection. We happen by great good fortune to have a peerless preparator, one who, given a reasonably complete skeleton, can mount it so that it is a joy to behold.

To be sure there are a good many odds and ends in our halls of vertebrate paleontology which are not very inspiring to look upon, but they are safe there now and they will gradually be weeded out as finer specimens are found to replace them. Obviously the scientific name which was originally bestowed on these creatures has to be a feature of the label, along with the name in present use, if any change has had to be made. Details are added concerning the horizon, exact locality, etc. In most cases we have tried hard to add some descriptive material concerning the affinities and habits of the animal in life, so that the labeling in this department has many redeeming features, but the point is that the individual specimens here are a pleasure to look at, and again I seldom see a visitor reading a label.

Let us consider how many hours I and my colleagues have spent in painstakingly looking up the name in most recent usage to put on some individual specimen of bird or mammal, fish, reptile, or invertebrate. I know well that the time thus spent represents weeks or months in the aggregate, and I know now that mighty few of these labels have ever been read at all. If a scientist walks through our halls, he knows what all the creatures are; apparently no one else cares.

We might try sorting over the whole collection and then see whether, with only arresting specimens preserved, each displayed with its label in large clear type, they are any more closely examined than is the miscellaneous assembly which we exhibit today.

There are animals which have attributes which might be pointed out to give them a real human interest. The hedgehog is immune to snake poison. Five or six different unrelated forms found in various parts of the world are all marked alike with black and white and all smell like skunks. The tiny Chihuahua dog and the St. Bernard are both descended from a Central

Asian wolf. We might show a native porcupine with some fine examples of Indian quill work. Examples could be multiplied indefinitely.

Some birds have inspired famous poets to write great poems. Others have some outstanding anatomical peculiarity. The claws on the elbows of the wings of one South American species called the Hoatzin, which at first sight looks like a pheasant, prove quite conclusively its reptilian affinities. A matter of this sort might well be stressed: show a mounted adult and a fledgling, in a flat jar of alcohol, with the wings stretched out to show the persisting claws. We have specimens, and if this does not make the fact we wish to emphasize clear, a sketch well enlarged will add to the educational value of the display. We have never made sufficient use of drawings and photographs to illustrate special features which we may wish to call to the public's attention.

It may be that, leaving the demonstration of zoölogical affinities to our synoptic hall, we could elsewhere display animals and birds so that the exhibit might awake imagination or answer questions in a way which would surprise many people.

The halls where the animals characteristic of the various faunal regions are shown are the ones where obviously popular interest has waned or completely disappeared. The public walks through there without even stopping. The fact that elephant and lion are found in Africa is at present well known and of no interest to anyone. Our African elephant is a somewhat runtish monstrosity, and there is no space available to show a really fine adult. There is little to incline one to pause to behold it.

It may be that one trouble with our disregarded labels is that we are trying to make them all things to all men. The intellectual levels represented by our museum visitors are certainly much more diverse than they were, let us say, one hundred years ago. Then, we learn, members of the so-called upper classes were supposedly the only ones who ever examined museum exhibits. I doubt personally whether fifty different sorts of turtles, all looking alike and each with an individual ticket giving its scientific name and the place whence it came, were examined with any more interest one hundred years ago than such a series is examined today. We have been proud of our profuse public exhibition of turtles and tortoises. Nevertheless, I do not believe that it is of any more interest to the public than it would be if it were composed of only a half dozen outstanding examples. Let us say that we displayed a giant tortoise from the Galapagos and one from the isles of the Indian Ocean, showing how difficult they are to tell apart, one from the other. Suppose we showed a typical sea turtle, a soft-shelled fresh water turtle, one or two of the well-known land and fresh-water tortoises and terrapin, perhaps a diamondback; and then nothing more but big open-faced labels telling why these were chosen for exhibit. Would this improve matters and capture more attention?

The exhibit which Ernest Dodge and I set up in the Museum at Salem, to show the hawksbill turtle and the uses made of its shell all over the world, is really stunning and has received a lot of attentive interest. Here again, however, the beauty of some of the objects which the Salem Museum had to display has surely helped.

It sounds as if I were growing pessimistic and skeptical about the whole matter of museum exhibition. I am afraid this is true. (Remember, that I am not talking of art museums, concerning which I know but little.)

When the Harvard Forestry models were on exhibition in the Botanical Museum and before they were moved to the Fisher Museum at Petersham, they really aroused enthusiasm. Visitors actually not only advised their friends to see them but returned to do so themselves. There was a reason. These models told a story and they told it well. Each individual unit carried an unbelievable illustration of reality. I was convinced then, and am now, that here was the answer to the whole question of groups for museums without great sums of money to spend. Small creatures to be sure might be difficult to display convincingly in this way, but a herd of elephants, for example, with African natives, shown to give scale, could be made most extraordinarily convincing because the fact could be conveyed and made realistic that elephants often travel in enormous bands, and no museum group, however expensive and elaborate, can show more than a single animal or a few at most.

Visual education from this time forward will come more and more to the fore. I mean education in the sense of movies with sound. These have one enormous advantage over most other forms of display, inasmuch as the visitor may rest, relax, and even remain seated while being instructed. There is no question in the world but that museum fatigue, both pedal and ocular, is one reason why strangers frequently rush through our halls.

I have said enough to show that the question of museum exhibition needs to be reëvaluated and restudied, and this not by the curators themselves. They are naturally the ones who shirk any responsibility for exhibits and who want to stick to their lasts in the research collections. The problem should be attacked by experts chosen from other walks of life.

Our task here in Cambridge is not the same as that to be attempted by one of the great metropolitan museums visited by tens of thousands from all the various social gradients of a great urban center, but the fact that it is simpler is no proof that we have come nearer to solving it. Our halls provide a certain amount of intellectual stimulation to groups of college students and to boys and girls from the private and public schools in our neighborhood when these visits are made in the company of competent teachers. If the teacher knows how to pick out a certain object and emphasize the story which it tells, interest can be awakened and the youthful visitors may really experience a first-class intellectual thrill. The average untutored visitor seems destined, however, to have a pretty slim time of it.

I have insomnia, as I have often remarked, and thus having occasions to think during the quiet hours of the night, I found myself, not long since, trying to sketch out a movie animation depicting the evolution of the horses. The gradual transformation of form in connection with changing habits and environment presented no unapproachable problem, so complete is the geological history of the horses. But try to add to this concept the spacial element, and one fetches up with difficulty at once. We know that there were at least two migrations of horses across Bering Sea, from the New World to the Old. Did the descendants of the first migrants die out, or did evolution continue in the Old World as it did in the New, and at the same rate? Did the horses finally, as we understand them now, go to the Old World in a later migration to mingle with individuals which they found there already in the same evolutionary stage of development which they had reached here, or did they simply supplant those animals which had passed earlier to the Old World and which then conveniently died out? In other words, are the wild asses, the wild horses, and the zebras of Asia and Africa descendants of the first migration only, the second migration only, or a mixture of both? Here the paleontological record is not clear, and the animation, I am sorry to say, is impossible.

After all the foregoing was written, I discussed this whole matter with my daughters and got a fresh and, I think, a worthwhile point of view. They maintain that there are two categories of museum visitors. First there are those who *have* to look at the exhibits. Those who go to the museum because they are taken there, their noses rubbed against the exhibits so to speak. They examine them because what they observe will aid them in passing a forthcoming examination. These may be termed the "involuntary observers."

The other come with a point of view which my daughters maintain has changed completely during the last few years. They claim that the dictum that a successful exhibit should consist of a series of labels illustrated with well chosen specimens is as dead as the dodo. They declare that systematic and synoptic collections of mounted animals have no popular appeal whatever, no matter how well they are done. An exhibit, to be popular with the average run of free-will visitors, must be one of two things. It must be arranged to answer questions which actually arise frequently – note that this applies to our collection of North American birds – or it must be so inherently beautiful or strange as to focus and hold the attention of the visitor.

They believe, and I am not at all sure that they are not right, that moving pictures and travelogues have made the natural scenery, the wild life, the appearance and customs of primitive people the whole world over, familiar to the man in the street in a way that has never been the case before. Exhibits must therefore either recall pleasurable experiences which visitors have had, or they must be arresting in novelty.

They point out that the well-mounted fossil animals are of this nature. So, also, are the birds' nests and the exhibit illustrating insect architecture. In a somewhat lesser degree, so is the Alexander Agassiz Memorial Room, by reason of the beauty of the models of the Pacific Islands and their timely interest. The brilliant colors of the denizens of the reefs, shown adjacent to the models, may justify the preservation of this room. The synoptic rooms of invertebrates and vertebrates must be maintained for what I have called the "involuntary visitors," those who *must* visit the Museum to obtain certain types of information.

The girls maintain that the geographical rooms are wholly outmoded and that naturally they are passed by with no heed. This, indeed, certainly corresponds with my own observations as I have watched the rather meagre public who visit our exhibits. Our Agassiz Museum suffers from the fact that it is inaccessibly situated and can never expect to be crowded except when there are special attractions to draw people from Boston and elsewhere. The perennial beauty of the glass flowers attracts several hundred thousand visitors a year, few of whom have much of any truck with the rest of the museum.

People, as I said before, swarmed to see the forestry models. They went just as they go to see the magnificent habitat groups in the museums in New York, Chicago, Philadelphia, Washington, and San Francisco. We, of course, have neither the space nor the means to make possible displays of this sort. It might be possible to create exhibitions which would compete successfully with these expensive groups. If the rooms in which we now have the displays of animals faunally arranged were darkened, artificially illuminated, and ventilated, and the same sort of well-executed miniature groups were installed about the wall that were shown to illustrate New England forestry, I believe that the popularity of the Museum would be increased and that the instructional value of the exhibits would justify the trouble and the expense of preparing them. The use of this sort of groups would enable a much more ample panorama to be depicted than is possible even when the actual mounted animals are used.

You could show such sights as are to be seen in the Game Reserves of Africa and convey some idea of the hundreds, or even thousands, of animals which occur there together. The effect of the great bird colonies scattered over the world could be conveyed in terms of three dimensions and, for some strange reason, everything shown in this way carries more conviction than any painting or photograph. Possibly – and mind you I say possibly, with due humility – if we had a hundred of such models, carefully chosen and beautifully executed, with comfortable seats from which they might be enjoyed by the spectators, is it not possible that we might have a thronged rather than a relatively empty Museum?

My statement of the thesis that beauty alone is a touchstone whereby the interest of the casual visitor may be tested may not be wholly correct, and I may be barking up the wrong tree. I say this simply because a few days ago I received a little pamphlet from the American Museum of Natural History in New York, entitled *Five Editorials*, by Albert E. Parr. In it he says,

> When science was very young the gap between scientist and layman was very narrow. It could be bridged by a general education which was still capable of embracing all available knowledge. And the scientist was a scholar in all subjects.
>
> As science advanced the gap widened. General

> knowledge fell to the bottom of the crevasse, and the layman became unable to reach across.
>
> We have come to realize that this divorce between expert knowledge and public comprehension is one of the severest obstacles to the further progress of civilization.

The situation he has pointed out, when you come to think of it, is entirely true. Then he adds:

> There is no general solution for the problem applicable to all disciplines of thought. It must be separately solved for each major branch of learning. Our question is how to achieve a solution for the sciences of natural history through the medium of the public museums. What we must find or create is a common ground on which scientific knowledge meets public experience.

I wonder, however, whether Parr is not really arguing with his point of view fixed on the visitor with sufficient intellectual curiosity to visit a natural-history museum in the same spirit that one would subscribe to a serious advanced extension course of a university. Personally, I believe this type of visitor to be a *rara avis*. The vast majority come to seek amusement or relaxation, or in winter often just to get in out of the wet or the cold of an inadequately heated urban apartment. I suspect that the best we can hope is that they will pick up some bit of information incidentally or by chance, and for this reason information must be presented in a sugar-coated form. I feel a real regret when I agree with Parr as he says,

> It would appear that the natural history museums themselves fell victim to the conceit of admiring nature more for its charms than for any power it might hold over man. A species which had lost all relations to man's affairs by becoming extinct was for that very reason treated with a reverence in museum display never accorded man's living enemies or friends. Even the subjects explained in the interpretive exhibits have almost invariably been chosen for their fascination rather than for their importance.*

Parr is, I am afraid, correct; nevertheless, personally I wish I did not have to abandon the notion that an exhibit which is very fascinating to look upon is by that fact alone very arresting to the public and from one point of view very important.

* The quoted passages are from editorials reprinted from *Natural History*, April and May 1943.

CHAPTER VIII

The Spice Isles Forty Years Ago

MY wife was just a little over twenty years of age when we visited Ternate. I was two years, two months, and two days older. This little yarn is not then a treatise on the natural history of the islands of the Great East, but it does record the impressions of two young Yankees when they were of a very impressionable age.

This tells of visits to that long-settled paradise of Amboina, the scene of the ancient labors of Rumphius, whose *Amboinsche Rareteit Kamer* (1705), is to be found in the library of the Museum of Comparative Zoölogy at Harvard and whose grave we visited. Its inscription was still legible in spite of the fact that it stands in what is now a high and lush forest of mighty trees in a climate which is very hot and very rainy.

We stopped at Ambon, as the island is now generally called, on our way to Papua and again on our return. We called also at various lovely little bays all around the shores of Halmahera, twice at the picturesque little city of Ternate, nestled in undescribable greenery along the shore of a charming bay and lying peacefully at the foot of a mighty and mildly active volcano of the same name. We called at ports on the south and north coast of Ceram, then wild and but little affected by Dutch influence, although there was a fort and a tiny garrison at Wahaai. We caught no more than a glimpse of the tiny isle of Obi Major.

I have gathered together the impressions and vivid memories engendered by these golden experiences. They have come to mind upon uncounted occasions, and the only sad feature I must admit is the thousands of times when I have harbored the hope of a possible return for a longer visit, in vain.

Anyone who may become imbued with curiosity at some scene which I have described or some story I have told can have his curiosity well satisfied by reading, as I have just done, a book called *Nusantara, a History of the East Indian Archipelago*, by Bernard H. M. Vlekke.* Vlekke makes the great days live again. He tells of the cruel and bitter struggles of Dutch, Portuguese, British, and Spaniards which bathed these idyllic shores, these isles of peace and superb natural beauty, with the blood not only of the avaricious contenders for the coveted spices which the country produced, but with the blood, sweat, and I doubt not, the tears of the simple souls whose homes they were.

So the motive to set these notes down came forward this Sunday morning after the passing of nearly forty years. The headlines I read in the paper today, Aug-

* Harvard University Press, 1943.

ust 13, 1944, were "Airdrome on Halmahera Blasted." "Planes drawn up in Formation on Jap Airfield near Sorong in Dutch New Guinea Bombed." All this is hard to believe, and what is sad is that all the fierce strife just past and still to come will cruelly punish the myriads of simple souls who called these places home and who have learned the arts of peace and achieved a real prosperity from the Dutch, who have metamorphosed themselves. For with the passing of time they have changed over from conquerors looking at everything with covetous eyes to kindly guardians and true teachers who at long last were giving their wards the most intelligent and understanding rule dealt out by any of the colonial powers on earth.

My diary reads:

> The sea gardens are a delight pure and simple. Incomparably finer and more varied in coral forms than those at Nassau, but apparently the alcyonaria are less abundant and varied. This is a most interesting town as an example of Colonial administration; for an oriental town it is most perfectly clean and neat. The shore collecting here is very good and we got some echini (sea urchins) as well as opuiroids (brittle stars) and some starfishes. The passer ikan (fish market) here is small, a little disappointing.
>
> We walked to a cave several miles from the town, called Batu Lobang, and found bats of two species

therein. I think there was one more species which we did not get. Ah Wu saw a sea snake swim past our ship but he did not manage to catch it, greatly to our disgust. Birds of all sorts seem very scarce in the woods back of town.

Piru, South coast of Ceram

A beautiful sheltered sail from Ambon to Piru at the head of Piru Bay. Found a small town with only two European officials. A tame cassowary of the local species walked familiarly in the Kampong. Went off at once upon landing into the sago woods and got some butterflies, beetles, also a few lizards. Then we walked out on a road westward from the village, through cultivated fruit trees and little gardens, and soon came to a small river, on the bank a Varanus and a fine Lophura were shot.

The very peculiar lizard which I then called Lophura and for which an earlier name, Hydrosaurus, has been found, was found to be common on Ceram and Halmahera. It looked precisely like one of the great prehistoric fin-backed lizards in miniature. It was about three feet long, a rusty brown in color, and with a great "fin" on its back, which stood up at least four inches in height and six or seven inches long. The beast we shot here was in a tree and we frightened it before we ourselves saw it, and down it came flopping into the stream and swam quickly ashore. Shot, preserved, and studied in Cambridge, it turned out to be a very distinct new species which I named *Hydrosaurus weberi*, after Professor Max Weber of Amsterdam, who had given us much good advice regarding our trip. The first member of the genus described was from Ambon, *Hydrosaurus ambonensis*.

> Next a fine blue kingfisher and some other birds were bagged. Here we saw gorgeous parrots and lories everywhere and also that large white cockatoo with the rose crest, but they "used about in" the tops of such high trees that our guns could not touch them, but Ros bought one as a pet which is as friendly and tame as a puppy dog. At this moment it is lying on its back in her lap, its stomach being scratched.
>
> A fine place for butterflies. We papered fifty-five during the A.M. We certainly look forward to our return here, when we hope to stay longer and do lots more. A big black and white butterfly flies very slowly and is a perfect fool, so easily caught. We saw Ornithoptera butterflies here too, but all flying very high indeed. They are a glorious sight, the species is I think *O. priamus*.

This describes a perfectly typical day which, as the reader will see at once, was a mixture of modest but thrilling triumphs and bitter passing disappointments. As a matter of fact, we accumulated so much cargo during the voyage, as well as a very sick Dutch soldier at our next port of call, that we sailed on our return straight to Ambon from Wahaai, Ceram. My journal of the outward voyage reads:

> We reached here early in the P.M., after leaving Piru. Wallace himself had been to this very spot which, of course, made it all the more interesting to us. We shot a fine parrot as soon as we landed near the village, but had bad luck with several other birds, which after being shot fell into such thorny thickets that it was quite impossible to retrieve them. It was 92° in the shade here when we landed, and very damp. However, we caught a cerambycid beetle just like the one Wallace figures on his page of illustrations of Moluccan beetles. (We had gotten *Euchirus longimanus* which he also figured, at Ambon.) At the fort the Dutch doctor showed us many specimens, which we could not get as he was collecting them for a museum in Holland. However, he found us a man who, he says, is a good collector, and he is to gather material for us to pick up on the return voyage. The school here is held in a little church and an Ambon Christian teaches the little Alfoer kids, who sing very well indeed.

We found a snake the morning after our arrival which was like nothing which I had ever seen but which did not *look* poisonous. We treated it rather cavalierly and put it in a jar of alcohol. Examined in Cambridge, it turned out to be the celebrated Death Adder, one of the most venomous species in the world.

I may add also that on our return visit we found that the man whom we had taken on after his being recommended by the Doctor had a very nice little lot of snakes and lizards awaiting us. No more Death Adders, however. I strongly suspect that he had no yen for catching them.

Gane, Halmahera

Our first glimpse of Halmahera, also called Gilolo. It is on the southwestern coast of the island. This is a tiny and very lovely spot. We hear that in a short time steamers will not stop here any more. Little cargo seems to be forthcoming.

Lots of birds everywhere. Lories, licking great red flowers with their brush tongues. Ros did good work with small insects, and Ah Wu got some fine lizards. There are no open paths about here, so we find trouble moving about to collect. I took a boat later in the day and went over across the bay and tried to shoot some very large hornbills and white cockatoos. Wounded one of the latter but did not get it. Met some Alfoers here who were not specially hospitable so we returned.

I may say that many of these coastal villages are inhabited by Malays who have wandered in and settled. The true natives, who usually but not always live inland and away from the Malay settlements, are called Alfuros by the Portuguese or Alfoers by the Dutch.

Galela, Halmahera

Arrived about 12 M. After a very rough night from Ternate (of which more later.) At 6 A.M. we stopped for two hours at Soeppoe (or Supu) off the north end of the island and put off some 100 bags of rice. A very heavy swell made this affair a most painful proceeding. I was sleeping on deck, rather soundly, and awoke having rolled right against the heels of our beef supply, hitched in a row to the rail, uncleanly, but luckily there was no kicking.

We passed Morotai Island, high and very well wooded, surprised to note that it was larger than Batchian, which is 50 miles long. Had tiffin and then went ashore.

Shot a lory quite near the landing, then took a path into the interior which led to "The Lake." Along this jungly tunnel of a highway we passed through very interesting Alfuro villages. A booming sound like the beating of a gong caused us to look up into the trees, and I shot two large black birds but we only got one, the other fell in such thick and thorny stuff. No signs of *Semioptera wallacei.* Bagged a white cockatoo and then got two more lories of the same species. Just before we reached the lake we passed some trees with very small yellow blossoms. Here at last were the Ornithopteras. This species is velvet black with golden-yellow wing markings. They are very shy and for butterflies flew very fast. They are as easily frightened into flight as a bird would be. After many failures, we bagged five and found that we had no papers large enough to fold them up in. So I had the pleasure of carrying them all the way back to the ship. However, we got them here in about as good shape as they were in when they were caught. Our excitement at catching them I think exceeded that which Wallace has described for himself.

Our disappointment was equally great, for one example of a far finer species would not wait for us and flew from the tree where it was feeding. We also lost two other *Ornithoptera helena*, of which many more males are seen than females. We also took some fine Longicorn beetles; they also are shy and were really hard to take with our net after one had been located sitting on some leaf which is lit by a sunbeam in the forest. Mr. De Jongh, the Dutch controlleur here, is trying to get us a good collector and have some specimens awaiting us on our return here. This is the best place yet!

On our return to Galela it rained all the time and so hard that we couldn't go ashore. This was very disappointing. Mr. De Jongh's collector had a fine lot of insects, but not much else.

Tobello Island, off Halmahera

Obviously a good place to collect, but here it rains all the time. The Alfoers of this part of Halmahera are the most interesting of the island. Their houses are well carved with designs and figures at the ends of the ridgepoles. Burial customs are most peculiar. The dead they place in small houses set up on poles in their very dooryards. Here the corpse remains for varying lengths of time, anywhere from three months to a year. After this the bones are carefully cleaned and buried, or else they are kept permanently in the house in the case of some special favorite. The little buildings of the dead are covered with a mat roof, often of very good design. Commandant De Jongh also took us about here and very generously gave me some ethnographic objects made by the people of Kau, a town quite near. One was a shield with the blood of its owner on it; there were two spears, two winnowing baskets, and some other baskets, also fans, fire tongs, rice baskets, the things which the bride brings to her husband as samples of the sort of work which she is capable of doing. The Alfuro hat is very varied as regards shape and design and coloring.

On our return stop the weather was threatening, our boys not well (malaria), so did not send them ashore. Took a long walk through the interesting and very neat Alfurotown. Saw more parrots and lories than we have seen anywhere else; also some very large hornbills and many other birds.

The curious little houses of the dead we photographed again, though the light was far from good, and they are sometimes a bit aromatic if you get too near them. These Alfuros of Halmahera are very fine looking people. They wear their wavy hair long and twisted into a big loose knot over the left ear. Many have heavy drooping mustaches and beards. Generally they are very hairy and very well set up. They are extremely shy about having their pictures taken. Quite unlike the Sassaks of Lombok who like being asked to pose. We got quite a nice painted palm mat here.

Ake Selaka, Halmahera. Gulf of Kau

We were most hospitably treated here by none other than Mr. van Duivenboden, the very son of the owner of the *Hester Helena* which took Wallace from Ternate to Doreh Bay. There is nothing here but his trading station where dammar, etc., is collected from the surrounding country for shipment. He told us that a large Pitta was common here, also Wallace's standard wing bird of paradise. He sent a boy out with us and we heard several but did not shoot any.

Some large hornbills passed overhead making a noise like a puffing locomotive, which they do with their wings as they fly. On our return we are to spend more time here, and I shall try to shoot some. Mr. van Duivenboden said he would have a half dozen skins of Wallace's bird of paradise prepared for us by his natives by the time we return. Sorry I will not be able to get examples from Batchian too. The birds from the two islands are distinct, I feel sure. We found some mud wasps' nests here, many with spiders imprisoned within, also the wasps' eggs, pupae, larvae, and some almost adult specimens. The spiders were very good ones.

On the return trip we arrived early in the morning, and a very hard rain began. Nevertheless, I went off into the woods with a guide loaned to me by Mr. van Duivenboden. After a long, very wet, and generally unpleasant walk we arrived at a group of tall trees where the "burong burong palet" were said to congregate. The female bird of paradise makes a call very like that made by sucking against the palm of the hand – a sort of clucking (not a shrill note). We mimicked this for some time and finally males began to answer with a short song, really quite thrush-like, not at all like the wark-wok-wok of the regular birds of paradise. They are extremely active, jumping about constantly, acting rather like a big overgrown chickadee. I watched them for some time up among the half-bare branches of a very tall tree; finally they came nearer and nearer in lower trees, but of which unfortunately, the foliage was very thick. I shot and hit one. I saw five in all. We heard a number of others, both males and females. They are evidently common enough, but from what I hear they are very local. After I shot once they became shy and I did not get another chance. After waiting about for some time admiring the strange pitcher plants (Nepenthes), we came back to Mr. van Duivenboden's house, the rain having let up, where I found Ros on his verandah.

Mr. van Duivenboden had made up for me three skins of the great Pitta – I only got a glimpse of one in the forest, and also two skins of a large black sort of cuckoo, perhaps Centropus. I have it from Weeda and perhaps another species from Papua. Mr. van D. has a pet parrot intermediate in plumage between that of the male and female of *Eclectus pectoralis*. Enormous pigeons were common here, called in Malay "burong kum kum" from their booming call – a good onomatopoetic name. Some of these we shot and made up the skins, others we ate for a pleasant change.

Patani, Halmahera.

This is a small town with a mixed population composed of Chinese traders, a few Arabs, Bugis from Makassar in Celebes, various coast Malays, a few Halmahera Alfuros, and some Papuans. On the trunk of a large fallen tree in the woods nearby we found a very fine congregation of Longicorn beetles of various sorts. Our collecting party, to-day, was composed of Ros and me, a full-blooded Papuan volunteer (from Djamna), Bandoung, Indit, and Ah Wu. We shot a hawk and a small flycatcher – also a fine white cockatoo. This when shot, alas, had a ring on its foot, and a small chain hanging thereto. As we were carrying it home to the ship its owner appeared. He was a Malay gentleman and had been sent by the local radjah, whose henchman he was. Rather wrathily he asked us why we had killed the bird. We replied that we had shot it in the top of a tall tree in the woods. Finally after much palaver we passed out a roupea, but he insisted on a ringgit. We tossed the roupea on the ground in front of him and he walked off fuming. The bird was worth about seventy-five cents (three-quarters of a rupee; a ringgit was $2.50).

In a Chinaman's shop we got six skins of *Semioptera wallacei* taken near here by natives. He had some other bird skins, but nothing of special interest. We may buy some more when we get back.

[On the return voyage we] arrived here just before tiffin time so did not go ashore until about two. It was fearfully hot, and Bandoung soon came back knocked out. The rest of us went on to where before we had such good luck with the Longicorn beetles, etc. But the ground had been burnt over and beetles were scarce. We got only what I expect are common species. Indit got one splendid buprestid. We saw more, but as with some of the fine Longicorns they were too shy and flew off at once to high trees. Got a few bird skins which the Chinaman had made up while we were in Papua. A great Pitta, a Tanysyptera from which, like a fool, he had cut the legs, a 'burong mas' gold bird as the Nicobar pigeon is called, and a good *S. wallacei* in an interesting plumage."

Weeda, Halmahera

Arrived about mid-day and had tiffin on board, then we hired a native prau and went up the kali (river). We landed two Varani at one shot and we could have killed any number of others and also Lophuras. These great fin-back lizards simply swarmed even to the tops of tall trees. Put up a deer, but by bad luck did not get a shot at him. Saw a wonderful lot of birds – cockatoos, parrots, lories, also a very large species of lory, and some fine pigeons. Saw sign of "alligators" but did not actually see any. Shot a large black cuckoo, a crow with brown wings, a large hornbill, and a splendid big Carpophaga pigeon, I think almost the handsomest bird I have shot yet.

The river here was really a glory to behold and was navigable perhaps four or five miles. It is very narrow, winding between banks covered with great, overarching trees. Lots of fine tall Pandanus in fruit – the fruit look like great overgrown custard apples. In places the banks were very steep and rocky, overgrown with Nepenthes or pitcher plants, and on the top of the banks great forest trees 150–200 feet high of the most intense green and of beautiful shapes. About some hanging vines great numbers of what our Javanese boys call in Malay the "kupu kupu putih besar" were flitting. We caught a lot of them and they seem the same as what we got in Halmahera, and they are all decidedly smaller than the ones we got in Ceram, also different from the ones taken at Sorong Island. We have really a fine series of these big white butterflies, and some day someone may have a good time making a comparative study of these butterflies from the different localities.

I caught a medusa jellyfish some distance up the

river; its "flappers" all came off except one when I lifted it out of the water. It was ashy gray with leprous white pustule-like markings. The radial canals and "flappers" are sky blue; i.e., the long flappers, the short oral lappets were gray, the same color as the body.

Lawoei, Obi Major Island

Arrived about twelve, had tiffin, then went ashore, blowing very hard and very rough. The only white man in the place is coming with us to Ambon to go on to Fak Fak in New Guinea, where he was stationed once before for two years. He tells me that the "kupu kupu radja" the Birdwing or Ornithoptera butterfly, is usually very common. I think *O. croesus*, as he said it looked to him like the one on Batchian, which Island is type locality of *O. croesus*. We did practically nothing here. I am sure it would prove a fine collecting ground indeed. This island was originally uninhabited, so the people here are all immigrants, mostly the usual types of coast Malays, but there are a few families of Alfuros from Halmahera. People who cling very tenaciously to their own manners and customs. Following e. g. the same burial customs as in Tobello, which is the region whence they came.

These little islands, scattered along the east coast of Halmahera, are those which were formerly called the

Moluccas or the Spice Islands, a term now used for islands over a much larger area. Two of the original Moluccas once had great importance. One is Tidore, the glory of which has now in great degree faded and the town become of negligible importance commercially. The other, Ternate, has not decayed to the same degree.

Both islands are really volcanoes with a limited fringing area of lowland. In the town of Ternate, directly back of the beach, the European section is very limited in area, but is composed of well-built houses on regular streets well laid out. Beside this section, the Sultan's palace within its enclosure is a large rambling structure, most of the buildings built of masonry sides with thatched roofs. The bodyguard of the Sultan were a motley crew, wearing heavy broadcloth uniforms of deep blue which looked very much as if they had been made in Europe and perhaps purchased at second hand. The most prominent structure was the mosque or "messegit," surmounted by a very high pyramidal structure consisting of six roofs one above the other and each one separated from the next by an air space about four feet high. Beginning with the upper one, each roof was smaller than the one next beneath it, thus producing a curious effect as of a pyramid of steps.

This general type of architecture for a mosque is not uncommon about Mahometan Malay villages. The roofs are usually in two or three layers, each

UPPER LEFT: THE AUTHOR COLLECTING IN CERAM

UPPER RIGHT: A SASSAK OF APANAM, LOMBOK

LOWER: THE SOMEWHAT MOTLEY BODYGUARD OF THE SULTAN OF TERNATE

LOUISA BOWDITCH BARBOUR AND THE GREAT TORTOISE IN THE GOVERNMENT HOUSE GROUNDS, JAMESTOWN, ST. HELENA

smaller and above the other. None seen elsewhere had roofs as widespreading or numerous as this mosque at Ternate. As an example of Malayan architecture it was really, considering the primitive tools in the people's hands, a very stupendous and outstanding structure. I wonder whether it is standing there today?

It was by inheritance of claims made by the Sultans of Tidore and Ternate that the Dutch acquired suzerainty over all that part of New Guinea which they now possess. The western two-thirds, roughly speaking, of the mainland of Papua itself is claimed by Holland on the basis of these old feudal rights, based as they were some on slave trading and commerce, mostly about the Arfak region of Papua, and very tenuous at that. Waigeu, Batanta, Salawatti, and Misool are also included in the area acquired in this way. Here Malayan influence was felt considerably more strongly than it was in Papua itself.

Ternate is a trim little town, comparable to Ambon, only there are fewer European inhabitants. The fauna is very rich, but it is dry and we see hardly any butterflies. We have a splendid lot of beetles, especially weevils and Longicorns. There is a pond on the side of the volcano about four miles from town, and here in the long grass around the shores we saw Varani scutting ahead of us as we walked along, simply in herds. The lizards consisted of several members of the genus Varanus consorting together, the commonest one, black with many fine yellow dots. Each lizard was about a yard or more long, and as they raced off in droves through the high dry grass they made a most surprising amount of noise. I have never seen such a gathering of big lizards and wonder upon what such a population of large beasts can find to feed.

We saw also a number of the lizards with the high and long crests on their back and tails, probably the same species of 'Lophura' we first got at Piru in Ceram. I shot one Varanus with my last cartridge and some other water-haunting lizards with my little gun, which, bad cess to it, misses fire and jams regularly and is a perfect nuisance. Shot also a cuckoo and some other birds. Located a colony of weaver birds at work over a stream in a very high tree. A black-and-white sea snake swam by the steamer. I got a boatman and went after it; caught it. Indit got a good lot of small bats here. I hope they will turn out interesting.

On our return I wrote,

We notice a good deal of difference between the faunal conditions when we were here before and now. The Eupholus, common then, is still so now. The Longicorns taken in such abundance then are now not seen at all. The Rhinoceros beetle is now abundant, much more so than before, and we see many more small brilliant species.

This is lovely country; as I write we are sailing away. On the one hand is Tidore, an absolutely perfect cone, the top in clouds. Opposite lies Ternate, its volcano smoking a lot today and its summit quite clear. The sky is overcast and a hazy mist covers all the hills of Halmahera and the distant Moluccas. The hills of greens and lavenders marking the location of patches of cultivated soil simply baffle description.

Our food on board the *Both* was excellent and well prepared, but perhaps a bit monotonous. We carried a coop for live poultry, picked up all along the way, and had brought on board at Bali about twenty head of cattle and their fodder. These beasts were hitched to the railing about the stern of the ship. We had for some time some prisoners of war taken at Atjeh in Sumatra, also chained to the railing. There was a canvas awning over the decks and both the cattle and the "orang nanti," or chain people, as the Malays called them, were quite comfortable and happy. The men went ashore somewhere along the line, I don't remember where, to work at road building.

The cattle had perforce to be killed and dressed on deck, and as this operation was usually performed while we were at sea, until Ros got accustomed to the mess she was a little disgusted.

As I have said elsewhere, we had no refrigerating apparatus on board. I remember she often hove a sigh of relief as the hose was gotten out and the inedible remnants of the beef critter were washed overboard, to the immense joy of the sharks that always seemed to be waiting for the feast. There is not much deck space on a ship of 1,300 tons, and we always seemed to have a surprising number of deck passengers, as the third-class were called, traveling from one port to another. There were times when we were hot and tired and wished that life on board had not been quite so confined; but take it all in all the European officers, who as I remember it were four in number, and the Malay crew were without exception intelligent, cooperative, and extremely friendly. I have made many voyages on many much more luxurious vessels with a far less congenial ship's company.

CHAPTER IX

The Heavenly Twins

IF there is a single pair of islands in all the wide world which has been more frequently talked about in connection with zoogeography and therefore perhaps more often mentioned in print than any other, it is Bali and Lombok. Between these islands Wallace proposed drawing that line which to this day bears his name. It was set up to delimit the Asiatic from the Australasian fauna.

We know, today, far more than Wallace did about the distribution of Indonesian and Papuasian animals, and we know that this line does not have the significance which Wallace and others assumed for it when he was writing about these matters fifty years ago. I don't think that I am mendacious when I say that I was one of the first, if not the very first, to point out that these two fundamentally different faunal areas are separated not by a line but by a wide zone wherein one finds gradually increasing a predominance of types derived from Papuasia as one passes from westward to eastward, and vice versa.

Just the same, the contrast between Bali and Lombok is very spectacular. The tiger, for instance, reaches Bali, and so also the bantang, one of the most beautiful of all the wild cattle, and further eastward they do not occur. You no sooner set foot on Lombok than you notice the fact that nearly every Sassak house has a tame cockatoo, and, if you are as fortunate as I was, you may see a flock of these native small white cockatoos decorating one of those great fruit trees that stand in clumps here and there about the ricefields of the lowlands. This island is as well cultivated as Bali and has a very considerable population, and the primeval forest inland was not accessible to casual visitors in 1907.

The inhabitants of Lombok are called Sassaks. They are considerably darker, somewhat shorter on the average and more stockily and strongly built, and the women especially are far less strikingly graceful and alluring to behold than are the Balinese. The latter wear gaily decorated sarongs of batik as do the Javanese and also usually have a batik "kain" folded about their heads, though this is not by any means invariably worn. The people of Lombok – and mind you now I am speaking of the time of our visit nearly forty years ago – were usually dressed in a rather nondescript variety of trade cloth; the women we saw were usually wearing black, one piece of cloth wrapped to form a sarong, the other thrown over their shoulders or wound around their waists.

From waist up Balinese and Sassak women wore nothing except, perhaps, a scarf which was not for purposes of covering, except when a cool breeze blew.

The men of substance in Lombok rode chunky, thick-set ponies about 13–2 hands high, while the Balinese walked. The women of Lombok trudged along the shaded pathways, their produce carried on their heads in baskets which were much more crudely and less elaborately made than those seen in Bali.

In Lombok, men of importance also dressed in Javanese style almost always, with a white "kain" wrapped about their heads and with a kris stuck crossways through the long, narrow cloth which was wrapped around their waists. The handle of the kris was always placed so as to be displayed sticking up behind the shoulder blades. (I wonder how many display a kris today.) The men and women of the lower classes, especially, in every way used to be so much less appealing to look upon, and it is quite easy to understand why Bali, not Lombok, should have been chosen by the Dutch for tourist development and protection as a sort of native reserve, and why foreign visitors were encouraged to spend their money to visit it, and not Lombok.

So much indeed has been written about the artistic nature of the Balinese, and the Island has received such publicity, that I do not propose to do more than recommend a single publication for those who may be interested. This is *Balinese Character, A Photographic Analysis*, by Gregory Bateson and Margaret Mead.*

* New York Academy of Science, 1942.

So, as I say, the contrast between the two islands, except in their appearance from the sea, is in every way as striking as their faunistic differences. Bali is inhabited by a race of people who one and all are artists to their fingertips; there all the arts and none of the sciences have been developed, and the arts are cultivated to very high degree. The curious characteristic architecture is familiar from the many well-illustrated books which have appeared in recent years. So also one may emphasize the excellence of the wood and stone carvings, the weaving, the painting, the marionette and puppet manufacture, the ritualistic and classical mythological dancing, to say nothing of the acting and the dancing of the people themselves. Proficiency in all these pursuits is so widespread that it is difficult to hold oneself from saying that it is universal. Nothing of the sort is the case in Lombok.

Both islands profited by the fact that they originally grew no spices and produced no gold. Hence they were unnoticed by the Dutch until in 1894 the poor Mohammedan inhabitants of Lombok implored the help of the Dutch Government against the Balinese, their Hindu overlords who oppressed them continually. The two populations were more or less continually at war until a Netherlands Expeditionary Force finished off these disorders once and for all.

Lombok was then for the first time brought under direct Dutch rule, although in a few years local self-government was restored. The Hindus of Bali – and this is not by any means an unsubstantiated fact – mild and gentle as they seem to us to be, were pretty ruthless slave traders. The point I want to emphasize is that neither island by good luck was subject to Dutch supervision, for neither attracted much notice, until what may be called the era of ruthlessness had passed. The Dutch, in common with the British and Portuguese, were hard and often cruel taskmasters during the first two and a half centuries after their conquest, and the change which came after 1860 was fundamental and praiseworthy and very fortunate for the people of their two lovely isles. Here again I refer the reader to *Nusantara, A History of the East Indian Archipelago* by Bernard H. M. Vlekke.

Bali is considerably the larger island of the pair and the more densely populated. It, with Little Nusa Penida, just off its south shore, measures 5,561 square kilometers, against its neighbor's 4,729. In 1930, when the last estimate – mind, I do not say census – was made, Bali had 1,101,393 inhabitants and Lombok 701,290. Between Java and Bali, the water is very shallow, most soundings showing about 18 and 19 fathoms. Between Bali and Lombok the water is much deeper but still it is shallow, as oceanic depths go, being generally about 320 fathoms.

There is no mountain, outside of New Guinea, in the Indonesian area as high as Mount Kinabalu in British North Borneo, which towers 13,698 feet above the great lowland forest. However, Mount Rindjani of Lombok is not far behind with its 11,300 feet of forested slopes, while Mount Agung, the highest peak in Bali, is about 9,500. These figures I hope and believe should be indication enough that the scenery hereabouts is striking and impressive in the highest degree.

We landed in Bali almost forty years ago at Buleleng, then the only port of entry and an open roadstead on the north coast. There were no good highways as there are now, except in the immediate neighborhood of the port, and permission could not readily be secured in those days to stroll about the island at will. We did, however, drive out in little pony carts and see those astounding temples at Sangsit and Singaradja with their elaborate stonecarvings, very much like the work on the temples in Central Java but much more overdecorated and with a rather degenerate feeling about the workmanship.

Our approach to the island that early morning of January 24, 1907, was unforgettable. We had left Soerabaya at four o'clock the previous afternoon. As we drew near to the open roadstead it was fortunately dead calm, for otherwise landing at this point is quite impossible. All along the horizon, far off and to the left, as we moved slowly toward shore, we could enjoy the breathtaking panorama of the east and of Java. The most conspicuous feature of the landscape was, of course, the cone of the giant volcano Smeru, which emitted at regular intervals a prodigious mushroom-formed and mountain-sized column of white steam. This towered away upward to become a great cumulus cloud, and then, as if it were snipped off at the base by a Gargantuan pair of shears, it drifted slowly away with the breeze until, at perfectly regular intervals, another cloud and then another arose and took its place. The volcano Merapi, one of two in Java of the same name, was also in plain sight, as well as the Bromo, which is even higher, being just about 10,000 feet. Right ahead lay the mountains of Bali, all covered with the most sumptuous tropical forest growth. The lowlands were covered everywhere with the paler green of the rice fields, glistening like moist jade in the early morning sunshine.

As we approached the beach the view of Lombok was gradually cut off, and although it is less than twenty-five miles away, it was not until the following afternoon, as we sailed away and as a short but heavy downfall of rain cleared away, that the summit of the stupendous cone of Mount Rindjani came into view.

We anchored off Ampanam about ten o'clock in the evening. Arising early the next morning, we took carts drawn by those stocky little ponies bred on this island or, more often, imported from the neighboring island of Sumbawa, and then drove inland to Mataram, the site of the old Kraton, or palace of the Sultan, who had been but recently deposed by the Dutch and a new incumbent installed. Along the roadside we were interested to see the throngs of women coming in to the market at Ampanam, much as in the early morning women may be seen streaming along, also carrying their wares on their heads, to the market at Port-au-Prince in Haiti.

In Lombok there is no noteworthy architecture, and no evidence of any artistic handicrafts to be seen in the bazaar. The people in their general appearance recall the Alfuros of Halmahera far more than they do the inhabitants of either Bali or Java. No two peoples separated by a narrow body of water could show less evidence of ever having known one another. The contrast is even more startling than that between Key West and Havana, and to this statement little in the way of emphasis can be added.

While we collected climbing perch from mangrove trees near the landing place at Ampanam in Lombok, it was not until our arrival at Macassar in Celebes that we saw these remarkable little fishes at their best. My journal notes:

> Macassar is really a fine little city. Its well-kept, shady streets and roads are all arched over with the most beautiful and luxuriant trees. Excellent European shops of all sorts may be seen everywhere. There is even a cinematograph theatre. The *passer ikan* (fish market) is here so extensive and beguiling that we spent much of our time there during the several days we have lain at dock in port, and be

assured we intend to do the same on our return trip.

The market bespeaks the industry, skill, and ingenuity of the Buginese fisherman to take so many sorts of fish of all kinds of habits. Still more it indicates the omnivorous taste of the wealthy Chinese community in the city. Rays and sharks of many species were very prominent in the stalls, and how we wished that we had several barrels more of alcohol than what we had at our disposal. As it was, we put up a number of small Selachians, which were not too big to get in our tank. Underneath the platform on which the fish are displayed the muddy ground swarms with a snake, Chersydrus, which evidently is caught when fishing is carried on with seines hauled in shallow water.

My notebook is full of sketches to indicate the garish patterns and the gaudy colors of some of the fish which we collected, and I may say that in collecting these here we really went to town.

Perhaps even more memorable were the number of occasions when we drove out to where there were some mud flats uncovered at low tide. Here we saw again many individuals of the quaint little climbing perch (Anabas) creeping about the mangrove roots. Of these most amusing little fish there are eight species known from the East Indies. They may be seen creeping about on the mud and up on the stilt-like roots of the mangrove trees looking like nothing so much as little gray mice. These little fish have a wonderful adaptation, inasmuch as the operculum which covers their gills is so tightly fitted to the side of the head that if they start out on their terrestrial peregrinations with their gills well saturated with water, air does not get in and dry them out. The gills are curiously labyrinthine masses of tissue, not at all like ordinary fish gills.

The climbing perch are nothing like as ubiquitous nor do they occur in such great hordes as the mud-hoppers with the jawbreaking name of *Periophthalmus kolreuteri.* Scattered all over the mud were groups of these tiny, hopping, pop-eyed gobies, fish acting in a perfectly unfishlike manner. After many opportunities to see them elsewhere, we found them quite comonplace, but the hordes today of the little gray devils hopping about over the mud or occasionally climbing up on the mangrove roots were certainly amusing to say the least, to any one observing them as we were for the first time.

We found several little runs of quite deep salt water which led up to where the foliage along some higher land was heavy and thick, and where the dense shady trees overhung the little bayous. Here every once in a while we exclaimed to one another with surprise at seeing little sprays of water shoot up about eighteen inches into the air. For some time we were quite mystified, then to our delight I found that we were being granted an opportunity to see the insect-catching operations of that marvelous little fish with the fascinating name of *Toxotes ejaculator*. You will recall that Toxotes is the Greek name for an archer, and ejaculator – well, that is just about the same in English as it is in Latin.

These little fish, of which we secured specimens that are now here in the Agassiz Museum, shoot a jet of water into the air and not unfrequently knock a bee or fly from a leaf or flower to the surface of the water, where it floats helplessly for just long enough to be grabbed by the little fish and swallowed. I am sure that there is not one of you who will ever read this who would not have been thrilled as we were at the chance to see these little piscine sharpshooters at work. And having made the observations, not also have enjoyed with us our ride in our little carriages back to town along the road covered by those great arching trees which lead to one of the loveliest, cleanest, and most attractive of the tropical towns of the whole world.

You have seen antimacassars on the backs of the drawing room chairs in your grandparents' houses. These little tidies protected the upholstery from the macassar oil with which our forebears anointed their curly locks. The name came from the cajaput oil shipped from Macassar, but produced on the neighboring isle of Buru, which is largely covered with the pale gray trunks, of the cajaput tree, in places growing so closely one to another as to color the landscape.

The Malay name for the tree is *caju putih* or white wood, hence the English *cajuput*, the name for the tree and the oil produced from *Melaleuca leucodendron*, now also grown as a windbreak in extreme southern Florida.

CHAPTER X

Mostly About Islands

MY family love Boston; this means everything in any wise remotely related to Boston, including the east wind and the winter climate, with a fervent adoration in which I do not share. I have done my traveling in winter, for the reason that Boston does not deal kindly with me at that time of year; but I have had plenty to do in Cambridge, hence I have always had to travel and collect hurriedly. To be sure, winter, the dry season, is in the American tropics the pleasantest time of the year in which to collect, but it is not by any means the best time.

The considerable collections which we made when we were in the East Indies during short visits to many localities were gathered because we found persons willing to help us in several places and because we dropped off two trained Javanese boys for a number of weeks at chosen villages. The fact that our ship stopped at most of the same ports on the return voyage from New Guinea to Java as she did on the outgoing voyage made this procedure possible.

When we were guests on Allison Armour's research yacht, *Utowana*, and there were other scientific activities demanding consideration as well as my zoölogical collecting, we worked out a rather definite procedure for each new locality. If the harbor where the *Utowana* came to anchor was uninhabited, there was nothing to do but scratch for ourselves. Luckily Rosamond has a fine sharp eye for a land mollusk, no matter how tiny, and often helped with collecting these. She does not shoot, so did not help with gathering birds or lizards. Many of these creatures were collected with a .22 rifle, the cartridges being loaded with dust shot. When, as was usually the case, we anchored off a village in the Bahamas or one of the islands about Haiti, we generally went ashore first to size up the population. You must remember that all people who met us for the first time were entirely convinced that we were crazy. I always carried a sack of small coins, British or Haitian, for the particular area of which I write. These we would display freely, and all and sundry would be informed that we would buy living creatures of the various groups of animals which we knew from long experience might reasonably be expected to be caught without doing the specimens too much damage. We would advise our helpers to roll stones over, and search under banana trash and driftwood, seeking out the little snakes and lizards that hide under such material.

We carried cans, jars, and canvas sacks of various sizes to lend out as container – and, I may add, the temptation to purloin these was often too great to withstand. Usually we picked out a youngster, either a boy or a bright young girl, who could head up the collectors. If they showed a willingness to scatter off into the brush and go to work right away, we ourselves would go to catching butterflies, collecting birds, or other tasks at which we knew they couldn't help. If, however, and in Haiti this was by no means seldom the case, they simply persisted in standing about to stare, there was nothing to do but go back on the yacht, telling them first that we would come ashore later in the day and buy what they had found. This, of course, was not what we most wanted to do, but the point was to get the largest amount of material in the shortest possible time.

In the Bahamas the natives are unbelievably poor, but in most cases they had seen yachts. In many places around the coast of Haiti, and especially on the little-visited islands off its shores, the people would far rather stall about the beach on the off chance that they might get a chance to feel your clothing or simply to listen to unfamiliar speech. I always missed Greenway at Haitian localities because his French was much more fluent than mine or that of any of my family. He caught on to the peculiarities of Haitian Creole French with amazing rapidity, and while I can peck along with all the various and sundry forms of Spanish, Creole French is really not my long suit.

If we spent, as we often did, several days in one port, the second day I would go out with small silver and recover by exchange as much of our copper coinage as possible for use on a subsequent occasion. I add here one word of advice for others who may some day have occasion to do this sort of thing. This is something which I learned in Java nearly forty years ago. It is essential to buy everything which is brought to one by natives unless the quarry represents something which in the beginning you have said very definitely you did not want. If you do not do this, your crew will think that you have not dealt fairly with them. They cannot tell perfect specimens from damaged specimens or realize that after several hundreds or thousands of a species of land shell, for example, have been brought in, you naturally don't want any more. When a laggard comes along with fifty additional specimens of some species you have found to be really common, the temptation is to say, "No, I don't want any more." To do this is a fatally bad practice.

It is well to pay two to three times the price originally offered for something which turns out to be really rare. Be careful, however, not to stress too much the searching for rarities when they are not reasonably easy to find. Discouragement often results, and children, especially, soon realize that they would have made more if they had stuck to hunting common stuff. I remember at Bannermantown on the island of Eleuthera and at several localities on Crooked Island we were able to interest the school teachers in our project sufficiently for them to be willing to dismiss their pupils, and we got really unbelievable collections in short order. To be sure, we dumped bushels of duplicate land shells overboard, to the immense enjoyment of the reef fishes swarming about under the shade of the yacht's hull.

This, I think, gives a fair idea of our mode of procedure. What it does not convey is the thrill that came to us with each dawn. The distance between collecting points was often such that we could leave one at dark, run along at half speed, and then lay to and slowly approach the next one after daylight came. Coral heads grow fast, and many of the charts in the West Indies were made a long time ago, so a slow, careful approach to any anchorage is indicated.

Day after day we would come on deck, see how far we were from shore, then eat a hurried or leisurely breakfast as circumstances indicated, and get back on deck to survey the scene of our day's labors. Let us suppose we are approaching land during a dead flat calm, as usually occurs in the tropics in the morning before the trade wind rises. Everywhere the water is so clear that the unbelievable colors of the coral reefs are plainly to be seen. It is frequently difficult to believe that there are 15 or 20, or even 50 feet of water under the ship's keel, so completely clear is the ocean water. The reef fish with their lovely colors, the pelagic forms living nearer the surface, dash about with incredible rapidity of motion, all to make a constant kaleidoscope, ever variable and never failing in its beauty.

Allison would sometimes hang a strong electric light after dark close to the surface of the water and below the gangway, and here one might sit entranced for hours, watching the animals attracted to the light — the tiny flying fishes, looking like brilliantly colored moths, fish of many sorts, and crustacea beyond count, but most incredible of all, the squids. They would come flapping their wavy filmy fins daintily, the prismatic change of their colors constant and unpredictable, each combination seeming more lovely than the suit which the beast had worn but a moment before. William Beebe and other writers, far more gifted at painting a picture with words than I am, have described these scenes. I can only add a line. If you ever have a chance to look down into a really first class coral reef, pray for good eyesight.

It is often interesting to observe how frequently one thing leads directly to another which is completely unexpected. When Dr. T. E. White and I motored north in the spring of 1942, we decided to stop at Chapel Hill, N. C., to find if possible a tiny fossil.

About a hundred years ago Dr. Ebenezer Emmons, writing on the geology of North Carolina, reported finding three tiny fossil jaws embodied in coal, or rather, in slate, from a coal mine not far from Chapel Hill. One, the specimen which he figured and described as Dromotherium, is still in the possession of Williams College, which is very obviously where it should not be. A second specimen is in safe keeping in the Academy of Natural Sciences in Philadelphia, and subsequent study has proved that this is the jaw of a different animal that has been called Microconodon. The third specimen was lost, and no other has ever been found. They are fossils stemming from the time when the transition from reptiles to mammals was taking place, and whether they really represent mammalian or reptilian remains has been a moot point from the date of their discovery until now. So much for our reason for going to Chapel Hill.

While Ted White was looking for the coal mine, which he found had been long abandoned and the spoil piles broken down and buried in vegetation, I visited my old friend Dr. Robert E. Coker and met his charming young assistant, Dr. William L. Engels by name. The latter had been studying the animal life on the coastal islands of North Carolina. He had by chance made what to me was one of the most interesting discoveries in recent years. The fauna of these islands is, as might be expected, very depauperate, and on one of them, Okracoke beach, mice abound, and a king snake. Now king snakes are common all over North America, and wherever they occur, they always feed only on other snakes. They seize their prey and wrap themselves about the victim, and when they have choked it to death, they proceed to swallow it. The king snakes on this islolated spot had no other species of snake to eat, and Engel's experiments at Chapel Hill showed that this snake, which we together have described as a new subspecies, had lost all interest in snakes as a diet, paid no attention to them when they were offered as food, but ate voraciously when mice were presented. Moreover, for some unexplainable reason, this reptile did not kill mice by the normal king-snake method and the method which is used by plenty of other species of snakes which feed on mice. Rather it used the method of killing employed by the black snake and its allies, where the mouse is seized in the mouth of the snake and then pressed against some such object as a stone or a tree trunk by a curve or loop of the snake's body, the snake's tail serving as a holdfast or fulcrum. The snake from Okracoke showed a number of divergencies from any of those found on the mainland, but nothing else as remarkable as this extraordinary modification of habits.

Dr. Engels also had discovered on Sam Windsor's Knob, another small island, a rat snake of the genus Elaphe which proved to be new. Engels was just about to enter military service, and he asked me to write up the formal descriptions of these two forms and publish them under our joint names, which I did. This, of course, naturally prevented me from naming one of them after him. A few months later, however, Dr. Coker sent me a very distinct water snake from Shackleford Banks, which is another one of these offshore islands, and with characteristic generosity suggested that I name this for Engels, which I did with great pleasure, so that the work which he did in the area will be lastingly marked.

The point of interest is that while these banks are by no means recently built up, there is no reason to believe that they were not once connected with the mainland. This was, let us say, ten thousand or more years ago, so that in these cases, at least, incipient specific differentiation in a small isolated population does not need a very long time geologically speaking to make its appearance.

During the period of maximum glaciation, as I have described elsewhere, during the last ice age, enough water was tied up in the form of ice so that the surface of the oceans was considerably lowered, perhaps as much as two hundred feet. This naturally brought about the connection of many land masses all over the globe which were separated by shallow water. Then, with the change of climate, the ice melted, the water was returned to the ocean, and the separation of the land was again effected. You will see that in this way connections which existed between England and France, Siberia and Alaska, Sumatra, Java, and the Malay Peninsula, Australia and New Guinea, are easily explained, as well as many other geographic changes in the West Indies.

After I had thought about this matter, it occurred to me how strongly analagous was the case of the fauna of the islands around the coast of Haiti. I have had a chance on several occasions to visit Beata, Saona, Isle Vache, La Gonava, and La Tortue. On some of these, reptiles had previously been collected, but not on all, and I gleaned an extraordinary number of novelties which I sent to Dr. Doris Cochran of the United States National Museum, who was writing a monograph on the herpetology of Hispaniola. This was an act of pure friendship, since I yearned to describe these products of my own bow and spear, and since for many years I had specialized on studying material from the West Indies. However, she did an exteremely good job, and I have never been sorry that I sent the material to her. These islands stand in the same relation to Haiti as the coastal islands do to North Carolina. Their last separation must have taken place at the same time and for the same cause. They have produced a great number, not only of distinct subspecies, but distinct, indeed conspicuously distinct, species as well. Here therefore is more evidence, if more evidence is needed, that, given isolation and not too large a population, speciation goes forward much faster than one might naturally expect it to occur. Cases such as these are the best indicators we have of the speed with which evolution works in nature. My trip to Chapel Hill, which I knew from past experience was sure to be a happy one, was far more than that and unexpectedly led to uncovering these facts so wholly worth while recording.

The Hawaiian Islands, taken together, constitute the largest of any of the extremely isolated land areas of the globe. Botanically and zoölogically they present many peculiar features, and their birds especially are unique for a number of different reasons. The Hawaiian group of islands has been in considerable part cleared for cultivation, but it is by no means one of those island groups which has suffered most sadly at the hand of man. There are still large areas of dense and relatively undisturbed forest. Nowhere, however, has so large a proportion of the peculiar species of birds disappeared with such rapidity.

The species which have disappeared and which were once common in Hawaii, according to the latest information available are something as follows: The two little rails are long since gone. The peculiar mountain goose and the crow are on the verge of extirpation. Of the four thrushes of the peculiar genus Phaeornis, two may still be in existence. Of the five members of the family Meliphagidae – which for want of a better name may be called the honey creepers – of Australian origin, all are probably extinct. Of the Drepanidae, of which I will have a good deal more to say later, out of forty-one species known to have occurred no less than fourteen are almost surely gone.

Years ago Outram Bangs and I made a pact that we would struggle unceasingly to secure every Hawaiian bird which ever came our way. This was not simple postage-stamp, pack-rat collecting, but it was because these drepanid birds offer the most outstanding example of what naturalists call adaptive radiation of any group of birds of which I know.

Looking back over an unknown period of time, we must presume that there were two fortuitous arrivals of flocks of birds in the Hawaiian Islands, which at the time they came were probably birdless. One of these immigrations was from Australia, which resulted in the development of but two autochthonous genera, one, Moho, with a species each on Hawaii, Oahu, Molokai, and Kauai; the other, a bird called Chaetoptila, which was discovered by Peale in the course of the voyage of the United States Exploring Expedition under Captain Wilkes. This bird disappeared at a very early date, and there are only some half-dozen specimens in existence. Peale, writing in 1848, says, "This rare species was obtained at the Island of Hawaii. It is very active and graceful in its motions, frequents the woody districts, and is disposed to be musical, having most of the habits of Meliphaga. They are generally found about those trees which are in flower."

This is one bird which we have never been able to secure and of which we have no hopes of ever getting an example. We have all the other four. Two and possibly three of these were regularly caught by the professional Hawaiian bird catchers, who, however, were accustomed to release their quarry after they had plucked the small bunches of yellow axillar feathers which were used in making the Hawaiian feather capes.

These robes were the most regal garments ever worn by man, and the most valuable, if consideration is taken of the time and labor involved in making them. The yellow feathers taken from the mohos are highly prized, but were not the ones sought in making the capes of the greatest, paramount chiefs. These were rather the little axillar tufts of feathers from one of the drepanine birds, called the mamo, *Drepanis pacifica.*

Dr. W. T. Brigham says of one great cape preserved in the Bernice Pauahi Bishop Museum in Honolulu:

> This magnificent cloak, made entirely of the feathers of the mamo (*Drepanis pacifica*), may well be placed at the head of the list, as it is not only in superb condition but so far as is known, is the only one of its kind in existence. It is the historical cloak once belonging to the great Kamehameha, and to the last days of the Hawaiian Monarchy it was used to decorate the throne on public occasions, long after it ceased to be worn as a robe of honor. When its fabrication began, neither records nor tradition clearly tell, but there can be little doubt that some of its feathers were plucked during the seventeenth century and the unfinished work ceased in the last quarter of the succeeding century. It is believed to

have belonged to the ancestors of the king Kalaniopuu who was king of Western Hawaii during Cook's visit, and from him the young Kamehameha inherited the insignia with his portion of the kingdom. The late J. J. Jarves, historian and art critic, in describing this cloak says:

"His Majesty Kauikeaouli has still in his possession the mamo or feather war-cloak of his father the celebrated Kamehameha. It was not completed until his reign, having occupied eight preceding ones in its fabrication. . . . A piece of nankeen, valued at one dollar and a half, was formerly the price of five feathers of this kind. By this estimate, the value of the cloak would equal that of the purest diamonds in the several European regalia, and including the price of the feathers, not less than a million of dollars worth of labor was expended upon it at the present rate of computing wages."

On the neck border are a few iiwi feathers, and the present border of purple velvet dates from the reign of Kalakaua. The length is 56 inches; front edges, 46 inches, width at base, 148 inches; weight, 6 pounds. The nae or net of olona is close, uniform, of a dozen horizontal strips with several triangular pieces, and in perfect condition. Given to the Bishop Museum by Legislative enactment.

It is to be supposed that the members of the family Drepanidae, of which there were no less than sixteen genera, arrived a very long time ago, probably from Central America. Diverse and unlike one another as many of these genera are, they are all derived from the same ancestors, as anatomical studies carried on by Dr. Hans Gadow of Cambridge, England, prove; and probably these ancestors were American honey creepers of the family Coeribidae. If the Drepanidae are not derived from that very group as we see it today, the ancestral forms of the Coeribidae were contemporaneous with the migration of the band which arrived in Hawaii and gave rise to the drepanids.

Once in the islands, these birds fairly burst into what might be called an evolutionary frenzy, and forms with heavy finch-like bills developed; others with enormous, long, slender, curved beaks, adapted for probing the calices of flowers. Others became parrot-like; others like crossbills. Others are nondescript little birds like finches and warblers, the great majority of all the species being olive to yellowish green. One developed a gaudy coat, and Ciridops was as gay as our Painted Bunting.

I believe, however, the most curious, and indeed entirely unique, combination of structure and habit occurred in the case of the genus Heterorhynchus, of which there were species on Hawaii, Oahu, Maui, and Kauai. Three of these have vanished to a point where they have not been observed for many years. In all four of the birds, the lower mandible of the bill was straight and very woodpecker-like, whereas the upper mandible was long, slender, and sharply decurved. By opening the mouth to its fullest extent, the little bird was able to chip and chisel with the lower mandible in the manner of a woodpecker. Then the upper could be brought into play, and the insect larva disclosed by the chipping could be extracted by the long, delicate probe of the upper jaw.

I think I have indicated enough to show the morphological as well as the ethnological interest attached to this most extraordinary family. And, by the way, if there were no other reason for suspecting their relationship one with the other, I may say that the birds have a strange, pungent, musky odor which persists for years and years, and which serves to characterize the skins even after they have been in a museum collection for a century or more, as some of ours have.

At one time we had a male Ciridops and either a young male or a female, it is impossible to say which; but in any case, the second specimen was unique. My predecessor in office was offered a specimen of *Drepanis pacifica* for either one of these birds, and he chose to exchange the unique uniform green specimen for the proferred mamo, preferring to keep the brilliantly colored male because of its unique and particularly characteristic color pattern. Some of us were pretty sad about the whole matter until the exchange was completed and we found out that we had done better than we had expected, for the mamo we re-

THE AUTHOR IN HIS BACK OFFICE

THE "EATERIA"

DR. HENRY B. BIGELOW, PROFESSOR HARLOW SHAPLEY, T.B., PROFESSOR PAUL H. BUCK, PROFESSOR E. D. MERRILL, MRS. W. E. SCHEVILL

ceived was a mummy and, should it be advisable ever to do so, a skeleton could be extracted, a dummy made, and the feathers or patches of skin, after they had been softened, could be removed and replaced. So skillful are the best of the taxidermists of the present day that it would be possible still to have a skin and make, also, a unique skeleton.

Moreover, when the history of the bird was fully determined, its sentimental value appeared. It had been a second specimen in the Vienna Museum, disposed of in the hard times after the last war. We later found that it was one of the original examples from which the species was described, being a cotype collected in 1806, formerly in the old Leverian Museum. We had made a very much better bargain than we knew. I cite this as one of the little incidents in connection with life in the Museum which add a zest to an existence which in reality needs none.

Mr. Henry W. Henshaw visited the Hawaiian Islands for some six years, from 1896 to 1902, and made a nice collection which contained many species which today have disappeared. This we secured. More than that, he wrote an excellent little book on the *Birds of the Hawaiian Islands*, which was published in Honolulu in 1902. He called attention to the difficulty of field studies in islands, which naturally are so densely forested and where the mountain ridges are so steep and the rainfall so heavy.

He points out that of the nests and eggs of Hawaiian

birds we know next to nothing. Forty years ago, when Henshaw was writing, he was already complaining bitterly of deforestation accelerated by the greatly increased number of cattle which browsed upon the tender young shrubs, vines, and undergrowth, thus destroying young trees and preventing their natural increase. Trees accustomed to a protective covering of mosses, lichens, ferns, and vines, which shield them from the sun and wind and keep them ever moist, die when this protection is removed or destroyed and cattle can chew it away. Henshaw is inclined to believe that for some reason or another birds had begun to diminish in numbers long before the coming of the white man, and for some biological cause. Indeed, it may be a fact that some forms had already disappeared, and hence that once there were more birds in Hawaii than we now know of.

It is mysterious that some birds which have gone do not show any extreme differentiation of structure. In the case of other species, it might be postulated that they had reached a point where superabundant growth-force had pushed some structure past the stage of usefulness to where it had become lethal.

Henshaw makes another interesting observation showing that some of these birds were even at the time of his visit only local in distribution. He cites particularly the thrush of Hawaii, a member of that little group of four species all of which are now very rare or extinct. The bird had adopted food habits which subjected it to little or no competition; it was a confirmed berry eater, with food everywhere abundant and in great variety. Nevertheless it was confined to one tract of woodland on Hawaii. The thrushes on Oahu had already been entirely extinct for nearly a hundred years.

With a few exceptions, each one of the native birds is confined to a single island. It is rare that a species occurs on two or three members of the group, and one wonders why these narrow, inter-island channels, varying from ten to thirty miles in width, the islands normally plainly to be seen from one another, should prove all-but-impassible barriers. Henshaw believes that "having once entered and become established on an island, however, the several species have readily adapted themselves to the new conditions which, though apparently differing but slightly, have yet proved sufficiently distinct to impress a number of the birds with new and, in some cases, markedly different specific or varietal characters."

Now, after twenty-five years, our search for Hawaiian birds is concluded. There are a few rarities in the hands of private collectors in the Islands, but, generally speaking, I feel quite sure that our Museum here has about as complete a collection as it is likely to obtain. Good luck attended our efforts, and we have now a representation of all the species of the Drepanidae except two. We have nineteen genera, forty-one species, and three hundred and six individuals, obtained from all sorts of sources. It has been a lasting satisfaction that we have been able to gather for those who follow us in the Museum a collection which would never have been gathered had we not started when we did. Our suite is one of the four best in the world, and I still hope that some day one of the little flightless rails, Penula, may turn up perhaps in the house of some old sea captain – one who mounted birds to while away the time and brought them home to put under a glass cover upon the mantel shelf. You cannot say that the betting is a dead certainty against our finding one, but it is a very, very long shot with the odds against us. Penula and Chaetoptila are the only two genera which we lack.

CHAPTER XI

Two Pleasant Memories

YEARS ago I heard Mr. Justice Holmes say that life was made up of big things and small. He likened the contrast to philosophy and gossip. I could not help thinking of the same contrast in regard to memories. Here I have set down two memories, long treasured, which came to me as a result of my life as a naturalist – curiously enough, I think, of events which, in themselves, were extremely inconsequential, but which nevertheless have given me great enjoyment as I have recalled them on innumerable occasions. The finding of these puny creatures, of which I am going to tell, bulks far larger in my repeated thoughts than, for instance, does the first occasion that I saw a lion in Africa, even though he was wooing his lady love – with, to be sure, far less ardor and passion, I must confess, than one associates with the amours of an alley cat!

Dendrobates

I wish I could give you a real picture of Taboga Island as it was when I first visited it. Beautiful and serene it had been since first it caught the eye of the first Conquistador who sailed the blue waters of

Panama Bay, and a charming little week-end resort it certainly was when Rosamond and I went there nearly forty years ago.

To be sure, life in the City of Panama was not ever very strenuous, but even so, there existed a little coterie of pleasant people who loved a still simpler life, loved the breezes that blew in from the placid Pacific, and, above all, always yearned for the chance to eat on its native heath one of the world-famous pines of Taboga. You know that Panamanians will defend the reputation of a Taboga pineapple against all comers, just as the Costa Ricans will the pines of Turrialba.

Our first chance to go to Taboga came in the form of an invitation from the American Minister to Panama, Herbert Squires. He had acquired automatically, back in the days when the United States Government owned very few residences for its diplomatic representatives, a charming old house which our Government took over from the old French Canal Company. This building was spacious and rather pleasing in a mild mid-Victorian way, but it was situated way down town in the oldest part of the City of Panama where the streets are very narrow and mighty little breeze circulates. One could well understand the pleasure with which the Squires looked forward, after a busy week of work and entertaining in the Capital, to the time when they took their launch and, usually joined by a group of neighbors, set forth for the

Island. A lot of people had week-end cottages along the coconut-shaded beach. These little shanties were pretty primitive affairs, and they would be looked down upon today in this sophisticated age. Nevertheless, the little beach colony was the scene of much pleasurable hospitality, or dining back and forth between the Gorgases and the Squires and the Federico Pezets — he was then the very popular Peruvian Minister in Panama, with a most amiable Belgian wife. This group was joined by a number of members of the first families of Panama who had frequented the simple litle resort for years unnumbered. The wines of Chile and Peru had to be carried out from town to grace the feasts, but beside the pines, native cooks on the Island prepared many a lordly dish of yellow rice smothering a little bivalve which was dug locally and known as a clam (in Spanish *almeja*), and was cooked with oil. Believe me, they were unbelievably sweet and tender.

I remember that on the first occasion of my going to Taboga, Mr. and Mrs. Squires and their daughters, Colonel (as he was then) and Mrs. Gorgas and their daughter Eileen, the Pezets, Rosamond and her cousin Mary Head, Doctor and Mrs. J. L. Bremer, and myself constituted the party. There were pleasant walks, — through the funny little town, by an old church and an old French convalescent hospital which was not used much after the American occupation, for it was found that Taboga was not free of malaria. This about covered the list of attractions, besides the chief one, which was, of course, the beach.

It is quite a sail from Balboa out to the Island. The route is partly sheltered on the one hand by the fortified islands near Fort Amador and then, on the other hand, by the Isle of Otoque, one and a half miles long, so that the trip, while of several hours duration, does not involve very prolonged suffering even when the sea is a bit rough, which is not by any means always the case. The Island itself is quite high and mountainous, with several pretty streams running down to its shores. At a guess, it is about one and a half miles wide by two and one half miles long. At the end of the trip to which I refer, for I made many others in subsequent years, I left the girls to get settled and then enjoy the beach, while Lewis Bremer and I skirmished around to see what we might see. Now curiously enough, after all these years – for remember, this trip took place in 1908 – two things stand out very sharply in my memory. Remember also, we were both quite new to the tropics in those days and easily impressionable.

I first recall that there was a colony of bats in one of the great sapote trees growing on the front lawn of the hospital. These trees have singularly closely growing dark green foliage and provide a wonderful shelter. The other picture which flashes to my mind's eye now is Dendrobates. I have never seen this beautiful little frog so abundant as it is along the banks of the stream near the village of Taboga. The species *Dendrobates tinctorius* occurs in an infinity of widely different color varieties in the various places where it lives, but the one found here is coal black, lustrous shiny black, with marbling of the most vivid metallic green, glittering wet-paint green — and the whole effect is totally different from that of any other amphibian which I know. I soon learned something about Dendrobates too; after catching the first specimens which I saw I rubbed my finger in my eye and thought I had put it out, for this devilish little frog secretes a poison which is unbelievably irritating and active. It is known that in some localities it was used as an arrow poison by the Indians. They simply tortured the poor little frog with the point of a burning stick and then rubbed the tip of the arrow on the slime which it exuded. In Brazil, the slime is used for another purpose. The Indians on the Amazon have found that the skin of the toad rubbed on the skin of parrots from which the feathers have been plucked makes all sorts of fantastic birds which find a ready sale. The parrot acquires yellow instead of green feathers after this treatment and thus strange "contrafeitos" are produced.

Miss H. M. Robinson, who for very many years has been my secretary, has just reminded me that when first she came to my office in the Museum of Comparative Zoölogy at Cambridge, I had just received some living specimens of these green and black

Dendrobates from Taboga Island. We had ours alive for some little time, but for years when I went to Washington, I was able to renew my acquaintance, for an old pupil, Doctor W. M. Mann, had a number which lived extraordinarily well in the beautiful Reptile House of the National Zoölogical Park, of which he is the director.

It is an interesting fact that this little frog abounds on this Island in such extraordinary numbers and yet on Ancon Hill on the mainland, perhaps five or six miles away in an airline, where environmental conditions seem to be quite similar, I do not remember ever seeing any specimens at all.

None of my memories of Taboga are as vivid as those of this first trip. I remember subsequently walking way inland through a fine deep tropical forest, not as luxuriant nor as varied, to be sure, as the jungles on Barro Colorado Island or in the Sambu River valley of Darien where the rainfall, of course, is far more abundant. Nevertheless, the woods of Taboga are sufficiently well developed to make welcome shade and stretched out as they are along the valleys of the streams of the Island, they provide a good collecting ground for the naturalist. I remember a number of years later going out again to investigate a theory that on the western coast of the Island there were sea caves which harbored bats, and that these bats were vampires. The depredations of vampire bats on livestock in pastures near the Pacific terminus of the Canal were many and unexplained, and for some reason they were supposed to have been made by bats which must have come from caves on the Island.

For the benefit of those who do not know about these strange little devils, I may say that their teeth are curiously modified to enable them to nip the skin of an animal so skillfully as to cause little pain, but then to draw a lot of blood on which the bat feeds. For years they were creatures of mystery around Panama. The Army lighted the stables and corrals where the Army horses and mules were kept, for the bats will only attack in complete darkness. Later, Doctor Herbert Clark found where the bats spent their days and caught some which he kept very successfully in captivity, feeding them on defibrated blood from the slaughter house. He found, indeed that they carried equine trypanomyasis, a crippling infirmity of horses; just as my friend, Fred Urich at the Imperial College of Tropical Agriculture at San Augustin in Trinidad found that the local vampires were the carriers of rabies. But our vampire hunts on Taboga were in vain. The colony of bats which lived for years in the old sapote trees in the hospital yard were not vampires, but were a fruit-eating species of the genus Artibeus and quite tolerant of daylight, which the vampire certainly is not.

I wish that my dream might come true, for only last night I imagined that I tasted *arroz con almejas* and that there was in my jaws a slice of that unbelievably succulent pineapple, a taste of which was the invariable concomitant of a trip to Taboga.

Halobates

Now, the scene of the other little story which I want to tell you is laid on the other side of the world, and after you have read it, I am sure you will agree how fortunate I am to have it to recall.

We entered Wooi Bay on a still, cloudless morning in 1907. Wooi Bay is on Jobi or Japen Island, which lies in Geelvink Bay, to the northeast of New Guinea. Forty years ago the people of Japen Island were noted for their unreliability, and we entered the Bay with some little trepidation! We came to anchor some distance offshore just as the sun "rose red as God's own head," in the language of the poet. At any rate, the day began to be blistering hot and no sign of a sea breeze had as yet arisen to move the still sultry air which was humid from the showers which had passed during the night.

There was only one mercantile passenger on our little ship, the S. S. *Both*, of the K.P.M. (so called to obviate repeating that Dutch jaw-breaker, the name of the company, Koniklijke Paketvaart Maatschappij). The ship's launch was lowered to take him ashore. He was going to see if there was any dammer gum offering. He had been to Wooi before with the idea of drumming up trade, but only a few times to be sure; no one in 1907 had ever been there many times. The principal difficulty in Japen Island trading was the fact that the people of all three villages we visited, Pom, Ansoes, and Wooi, set great store by the *pacheda*, and were interested in very little else in the way of trade goods. The *pacheda* is an armband made from a ring of shell which has been sawed out from a large member of the genus Conus. For one of these ornaments the natives were itching to barter, and upon one they placed a very high value. To digress, I may add that we went so far, after our return to Boston, as to see if it might be possible to have china replicas of these shells made by Mr. Richard Briggs and thus make large profits. Needless to say, this project never eventuated.

Our morning's plans were to ride ashore with Mr. Sedee and see what we could pick up in the form of drums, bows and arrows, shields, etc., for the Peabody Museum of Cambridge. Our native collectors knew too much about the reputation of the people of Japen Island to be willing to attempt any natural-history collecting here, and I cannot remember now whether they went along with us in the launch or not, but in any case, we set forth and began to steam about amongst the gigantic communal houses built like enormous turtle shells set up on stilts over the water. There were two or three rows of these, some quite near shore and others well out in the Bay. It was amusing as we cruised among them with our launch, to notice that the great outrigger canoes were always tied up at the front of the houses. This was where the men dawdled about, seeing what was to be seen and discussing what was going on, whereas the women and girls occupied the back porch nearer land. The story which we got, and I guess it was true, was that there was the usual bad feeling between the shore dwellers and the wild people inhabiting the mountains, which were high and very densely wooded. As happens anywhere in the islands under these circumstances, the mountain people not infrequently raid the shore villages on the chance of killing a few men for the sake of adding heads to their collections and catching a few pretty girls and young women to help with the work at home. The position which the women occupy made it certain that in the event of a raid they would receive the first arrows, then naturally scream and give the alarm, and the men would have a chance to get away. Such is the chivalry of the savage!

We did manage to get ourselves into a bit of a jam by beating one of our newly acquired drums. We did not know that a man had just died in one of the great communal houses, and custom demanded that silence be maintained. Of this I was naturally unaware when I proceeded to try out the drum. An immense amount of shouting and gesticulating began, and the native members of our little boat crew advised our immediate departure; so back to the ship we went, ultimately to learn that the object of the silence was that the spirit of the departed should not be disturbed, and, in this particular case, should not be given what might be considered by the spirit as an invitation to return! Apparently, the late lamented was not a very popular character.

Our little launch had no sooner left the ship's side, than I noticed insects skating about on the surface of the water, and it occurred to me at once that here at long last I was face to face with Halobates. I had read of Halobates and built up in my imagination a strange romantic sort of story, thinking over what I had read to myself. Halobates is a wingless bug, related to the water striders of our little ponds and streams. It is a delicate, fragile, little creature which lives sometimes near shore and sometimes hundreds of miles from land – the only large group of insects, for there are about twenty-five species in the genus, which in this way attempts life on the ocean wave. The span of their legs is somewhat over an inch. To me, it always seemed impossible that such frail little beasts could manage to exist in such a precarious environment, for Halobates when caught in a tow net dies easily, I can only suspect by drowning, whereas they live skating about over the face of the deep in all weather, with the greatest conceivable agility.

The most interesting account which I know of the breeding habits of these creatures is one written in William Beebe's peerless literary style in his book,

The Arcturus Adventure (1926). Beebe met with many Halobates in the waters about the Galapagos Islands. He found that they frequently deposited their eggs on the floating feathers of sea birds and on tiny bits of floating wood. The young of the insects, when they first hatched, he found were consumed by young fish. The adult insects, however, must be too dry and juiceless to be tempting, else they would have disappeared long ago from the face of the sea, for these fragile insects, while they are widely distributed over all the tropical waters of the world, are not by any means common animals anywhere, and the group is ill represented in entomological collections everywhere. Sir John Murray, who was naturalist on H.M.S. *Challenger*, wrote years ago that he suspected that the habits of Halobates were like those of its fresh-water allies, and that he had seen the insects congregated about any small recently dead animal, such as a fly floating on the surface, the attraction being the juices of the animal, which they obtained by piercing its integument with the aid of their mandibles and then sucking the fluid by means of their maxillae.

We found their agility to be quite astounding, and that they swam about chiefly by means of their long middle and hind legs, and when alarmed, can make progress by long jumps. We had quite a time catching a series with our butterfly net. The tibia and tarsus of the leg of these bugs is provided with a fringe of hairs which, of course, helps them with these rather acrobatic movements. The absence of elytra and wings is constant in all the species of the genus, and on this account naturalists formerly thought that all the specimens which had been found were immature and had not reached the adult condition. The discovery of their eggs, of course, proved this theory to be erroneous. There is still a vast amount to be learned by observation about these extraordinary bugs. This is not easy to accomplish, since they do not happen to abound in what might be called convenient localities for such studies. I was just thinking about the matter of their habits and how much there was still to be found out about them when it occurred to me that Halobates needed a Fabre. This happened to flash into my mind the other evening as I was again reading a book by William Beebe, the *Book of Naturalists*, which appeared in 1944. This is an anthology of skillfully chosen selections from the writings of naturalists. On page 213 of Beebe's book there is a touching tribute to the poor country school teacher, Jean Henri Fabre, born in 1823, who lived in the south of France until 1915 and wrote with such deep feeling and simple charm concerning wasps, bees, beetles, grasshoppers, and the like, and with such an uncanny insight into their habits. He wrote with such beauty of diction that William Morton Wheeler, who sat with me day after day in my little lunchroom at the Museum, often used to declare that as a recorder and observer Fabre was in a class entirely by himself. He said he knew of no other naturalist who had ever written more lovely prose. So you can see now the curious concatenation of my line of thought: of Wooi Bay, Halobates, and the pity that no Fabre had ever been where he could work on the habits of these marvelous little insects, which in many respects seem to me one of the most romantic organisms in the world. Certainly in the course of its daily existence it survives in the most perilous home which any insect has ever chosen.

The foregoing would have been the last word until 1926, for until then entomologists believed Halobates' pelagic life to be unique in the insect world. The other day, however, I happened to show my little note on this bug to my colleague, Dr. F. M. Carpenter, who told me about Pontomyia. This is a little fly which was found near Samoa and which also inhabits the wide-open spaces of the ocean. In a way, it is more remarkable even than Halobates, for it lives by breathing the air which is dissolved interstitially in the water just as a fish breathes, while Halobates breathes the atmosphere as we do. I wish I could tell you more about this insect, but as yet there is very little known to tell. Some day, however, you may be assured there will be a fascinating story to unfold.

It is a somewhat curious fact that considering all the cruising I have done in tropical waters within the last forty years, I have only once or twice seen sea skaters, other than in the Geelvink Bay area of New Guinea, where, although the only series I captured was at Wooi, I saw scattered individuals in quite a number of localities. I realize now that I probably made a mistake in not trying to gather more individuals, but at the time I was under the impression that there was only one species in the genus, and for that reason, once the series which is now in the Museum was collected, I thought that the matter could be crossed off the list for good. I am sorry now that I did as I did.

CHAPTER XII

Retrospect

WHEN we married and finally settled down, we took a house on Fisher Hill in Brookline, where my wife had been born and brought up. After the death of our oldest daughter, Martha Higginson Barbour, my father felt that it would be better to make a complete change, and we moved to Boston in 1914. There we have lived during winters ever since. This has presented certain advantages from my point of view and certain disadvantages. I have met and seen more of a good many persons with whom I should have been less likely to come into contact if I had lived in Cambridge. On the other hand, the reverse has certainly been true, and I have not had the opportunity to entertain and to associate with my colleagues to the extent which would have been possible had I lived in Cambridge.

I see the contrast evident in the case of the choice made by my daughter Mary B. When she married Dr. Alfred Kidder II, they moved to a little house quite near the Peabody Museum, with which he was associated and where, until the war broke, he was actively engaged in teaching anthropology. The result was that Mary B., as we call her, became employed by the museum and became an expert restorer of prehistoric pottery, and threw herself heartily and completely into the life of the circle in which her husband moved.

My life in Cambridge has presented a natural contrast to life at home in Boston. Many of the visitors whom I have welcomed to the Museum and who have lunched with me at the "Eateria" which I have described elsewhere,* have become warm friends and in many cases have been persons who would have had small appeal to my family, with natural interests so wholly different from my own. That this may work the other way is shown by the following incident, inasmuch as a person who during the early years of my directorship gave me much pleasure was a connection of my wife.

Rosamond's maternal grandmother was Sarah Rhea Higginson, and one of her first cousins was Major Henry Lee Higginson. He married Mr. Alexander Agassiz's sister, who thus became our cousin Ida. I was fond of her husband, but both Rosamond and I were more than fond of her, although during the last few years of her life she was much of an invalid and we saw little of her. She lived in Manchester in the summertime and often drove over to see us, who lived at Beverly Farms, and vice versa. We were in summers only two miles apart. Likewise she was al-

* *Naturalist at Large* (1943), p. 162.

ways interested in the Museum and was also a frequent visitor there.

I remember the last of these visits. She had grown very old and feeble, and at times she became a bit confused and surprised to find me sitting in the chair which she had always associated with her brother's occupancy. On the particular occasion to which I refer, it took her a second or two to realize that it was not her brother Alex but me with whom she was talking. Then she looked up, seeing the large engraving of Humboldt hanging near my desk, showing him sitting in his study with his books about him, and her face lit with that charming smile which everyone who knew her will remember, and she suddenly said to me, "Why, that is Mr. Humboldt." She continued, "I remember him so well. I sat in his lap in that very chair when I was a little girl. He was a great friend of Father's."

This seemed to me an unexpected linkage with the past comparable to the one which Mr. Justice Holmes, my wife's cousin Wendell, often told about in his old age: as a boy he was taken by his father Dr. O. W. Holmes and his mother to watch a parade pass along Beacon Street on the Hill. He saw there in an open landau, seated and receiving the applause of the crowd, three men who had fought in the Battle of Bunker Hill. Between their lives and his, the whole life of our nation was represented.

Mr. Lowell, when he was President of the College,

used to enjoy coming over to lunch at the "Eateria," and everyone was enchanted when he came, for he was one of the best conversationalists who ever lived. Moreover he often called up and suggested that he would like to bring persons who might be visiting him in Cambridge and who, he thought, would be interested in seeing the Museum or in meeting some members of the staff. I remember one day when he brought Sir Frederick Kenyon, Director of the British Museum in London, and Sir Henry Miers, who was head of the Museums Association in Great Britain at that time. Though neither of these two men could hold a candle to Mr. Lowell when it came to sparkling vivacity and breadth of learning, they were extremely welcome visitors.

Mr. Lowell came to my house in Boston when I had meetings of the Wednesday Evening Club, and occasionally to dine. The last time, I think, was when he sat after dinner for a long and memorable evening, looking at the films taken on our second African trip in 1935. Frequently I made it a point to see him in his own house on Marlborough Street. I stopped for a cup of tea and a cigar on I wouldn't dare to say how many occasions after my Cambridge day was over. We lived but a short distance apart, for my house is on the corner of Beacon and Clarendon Streets, but a couple of blocks away.

I have really lived two lives. I need not say that the Cambridge life has been the one of adventure and excitement, the one which has provided the excuse, if in very truth it has done so, to write these lines. I can hear my reader say, "What stuff and nonsense. How can there be any excitement in connection with T. B.'s life in Cambridge?"

Let me give you an example, one that happened not long ago. The Trustees of the Peabody Museum in Salem gave me permission at my own discretion to tear to pieces their hideous and utterly insignificant natural history collections and remodel them to be attractive and instructive. I found, however, that some of the worst of the junk had a sentimental value which could not be left out of consideration. So I saved some of the relics and cudgeled my brain for some use to which they might be put. One of the specimens to which I refer was a mounted King Penguin. Now if you searched throughout the museums of the world there is no possibility whatsoever that you would have been able to find another penguin so utterly hideous, so vilely mounted, as my Salem candidate for salvage. I thought to myself, "What does this bally thing show?" It suddenly occurred to me, "Unquestionably it shows what was considered a good specimen, well worthy of exhibition, a treasured relic indeed in the eyes of those who visited the Peabody Museum say 125 years ago."

This gave me an idea. I put it on exhibition and set beside it a magnificent Emperor Penguin which I wheedled out of my friend Dr. Alexander Wetmore,

Director of the United States National Museum in Washington. This penguin, taken on Byrd's last expedition to the South Pole, is the last word in lifelike artistry, an example of the incredible perfection which the art of taxidermy has reached today. The old penguin, labeled to set forth its history (it was the first bird of its kind ever to reach an American museum), with the other placed alongside, made an exhibit of taxidermy then and now which is certainly vivid, and one which has been eagerly sought out and examined by visitors ever since the new hall in the Salem Museum was opened.

I have set up other exhibits to show what were considered acceptable parlor ornaments in mid-Victorian times – examples of what seems to us the incredibly bad taste tolerated in the houses in which our forebears lived fifty to seventy-five years ago. I have carefully preserved some of the panels of mounted game and one case of mounted birds, which were odds and ends from all over the world – badly mounted, too, on the unbelievably hideous branches of an artificial bush in an ornate case. With this is displayed, on the floor of the case, a young rabbit, and still nearer the floor there is a framed portrait of the hunting dog which once belonged to the owner of this horror.

This is a vivid warning to all museum trustees or Directors to accept nothing, unless they are very sure they are justified in doing so, to which a proviso is attached that it shall be retained on exhibition forever. Nevertheless, it is quite surprising how under the urging of the spur of necessity one's ingenuity may be brought into play and material which is at first sight useless from every point of view may be forced to point a moral or adorn a tale.

LIBRARY
LOS ANGELES COUNTY MUSEUM
EXPOSITION PARK

INDEX

Index